AN ATLA

THE BREEDI

OF

SHROPSHIRE

Editorial Committee

Peter Deans   Jack Sankey
Leo Smith   John Tucker
Chris Whittles Colin Wright

Published by the Shropshire Ornithological Society 1992

First published in 1992 by Shropshire Ornithological Society.

British Library Cataloguing-in-Publication Data
A catalogue record for this book
is available from the British Library.

ISBN 0 951868 90 X

Printed in England by Livesey Ltd., Shrewsbury SY3 9EB.

CONTENTS

FOREWORD

by Professor David Bellamy

The meres and mosses, the Severn and the Teme, the Stiperstones and Clee hills, are all sites that I knew in my early researches in Shropshire. For this book every one has been surveyed, along with the rest of Shropshire — each dingle, drain and cwm has been searched by a volunteer force of almost 400 birdwatchers to produce this little gem of a book.

Hard on the heels of *The Ecological Flora of the Shropshire Region* and the more recent *Butterflies and Moths of Shropshire,* this breeding bird atlas fills one more gap in our knowledge of the wildlife of this increasingly well-researched county.

Above all this book again demonstrates the power and effectiveness of co-ordinated volunteer effort in providing a thoroughly good read and, at the same time, a source work for soundly based conservation amid the unrelenting pressures on our countryside, not least its birds.

The Shropshire Ornithological Society, and everyone associated with the project, can be proud of *An Atlas of the Breeding Birds of Shropshire.*

The birds say thank you too.

David Bellamy
Bedburn
July 1991

PREFACE

The Shropshire Breeding Bird Atlas is one of the most important nature conservation publications ever produced in the county and is the most detailed study to date of Shropshire's birds.

The Atlas is based on six years fieldwork between 1985 and 1990 and includes up-to-date distribution maps for nearly all of the 122 species of bird now known to breed in Shropshire. An introductory chapter describes the wide variety of habitats, and special computer-generated maps show the association of particular groups of species with each habitat. Commentary by some of the county's most knowledgable birdwatchers explains the distribution, and relates it to these habitats. Population levels are estimated for each species. The birds are illustrated by local artists.

A Breeding Bird Atlas for Shropshire is especially important. The county lies at the meeting point of the Welsh hills to the west and the Midland plain to the east, with their contrasting terrain, land use, habitats and wildlife. This geographical position also determines the climate, which is intermediate between that of the warm, dry south-east of England and the cooler, wetter regions to the north and west, with considerable variation between the uplands and lowlands. Thus Shropshire is on the edge of the range of many British birds and any changes in the future should manifest themselves early here.

The Atlas therefore provides a bench-mark against which all future trends can be monitored, so it will become a major reference book. Local Authorities, voluntary bodies and landowners, and anyone else concerned with planning, nature conservation and protection of important wildlife habitats, will use the Atlas for the foreseeable future. Habitat protection is especially important, as birds can only flourish if their breeding and feeding areas are safeguarded. The Atlas has highlighted which areas and habitats are particularly important, and has helped identify sites, species and land uses which require care in the future.

The publication of this Atlas is only possible thanks to the efforts of the fieldworkers, mostly members of the Shropshire Ornithological Society (SOS), many of whom spent a considerable amount of time over several years searching for the more elusive species; the ten Area Co-ordinators, who organised the fieldwork, checked the results, and wrote most of the species accounts; the four artists who drew the illustrations; the Shropshire Wildlife Trust (SWT), who allocated office space, officer and computer time over the whole life of the project for the development and subsequent use of the TETRAD™ program to store and check data, and produce the final maps; the volunteers who input data, checked it and read through the proofs; the various typists; the map designer; the designer and printing staff at Liveseys Limited; the individuals and bodies who contributed to the production costs, particularly English Nature; the Countryside Trust, who financed the fundraising and promotion costs; and the Atlas Sub-Committee of the SOS, who co-ordinated the whole project and produced the publication. They are all individually acknowledged in the appendices. Particular tribute is due to Colin Wright who, as chairman of the Atlas Sub-Committee, led the whole project from beginning to end, collected in the data every year, wrote many of the accounts, retyped most of the manuscript and prepared it for printing, and still found time to keep everyone else on their toes. All these people can be justly proud of their contribution to the Atlas, a major reference book for many years to come.

Publication has also opened up new horizons and opportunities. Further fieldwork is already underway to survey in detail the important sites that have now been identified, and the results and distribution maps pose as many questions as they resolve. Several issues worthy of additional research are listed in the summary and conclusions, and it is hoped that the Atlas will inspire further inquiries into the fascinating world of Shropshire's breeding birds.

Leo Smith
Editor

SHROPSHIRE AND ITS BIRD HABITATS

Shropshire is a county of contrasting and dramatic landscapes. The Shropshire Hills, designated as an Area of Outstanding Natural Beauty, dominate the west and south. The north and east comprise the edge of the Midland plain. A major river, the Severn, and its tributaries, drains virtually the whole area. This natural foundation is overlain by human land use — agriculture and forestry, mining and quarrying, drainage and water management, and urban development.

The resulting interaction has created very varied habitats, reflected in the wide diversity of breeding birds.

NATURAL FOUNDATIONS

Shropshire is the largest land-locked county in Britain, with an area of 3519 square kilometres (sq. km) or 1346 square miles. The natural regions are shown in Map 1.

Natural regions

The north-west uplands near Oswestry are an extension of the Clwyd hills. Up to 330m high, the Carboniferous limestone and sandstone scarps have steep west-facing and gentle east-facing slopes, deep valleys, and occasional limestone cliffs and quarries.

Clun Forest in the south-west is also an extension of the Welsh hills. A rolling plateau of Silurian sandstones and shales, and Old Red Sand-

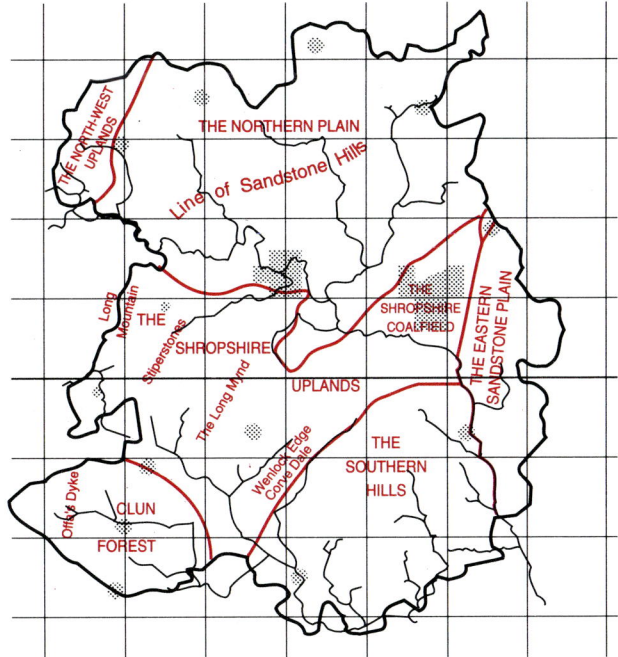

Map 1. The natural regions of Shropshire.
(adapted from *The Shropshire Landscape,* Trevor Rowley 1972, published by Hodder & Stoughton Ltd)

stone, it is around 460m high, and is cut deeply by the major valleys of the rivers Clun and Teme, and their tributaries.

Offa's Dyke, the historic Welsh border, traverses both these hilly western regions, and in the miles between them passes over Long Mountain, an outlying hill of the Shropshire uplands, which rise to over 500m and are based on Pre-Cambrian rocks, some of the oldest in Britain. The four long parallel ridges of the uplands — Stapeley Hill, the Stiperstones, The Long Mynd and the Stretton hills/Caer Caradoc/The Lawley/The Wrekin, together with the Silurian limestone escarpment of Wenlock Edge — all run south-west to north-east, and are separated by the valleys of the rivers West Onny and East Onny, the Stretton valley, and Ape Dale.

Where the uplands come to an end, in the north-east corner of this region north of the River Severn, is the only major industrial area. Previously the Shropshire

1

coalfield and the birthplace of the Industrial Revolution at Ironbridge, it is now the site of the sprawling Telford New Town.

Corve Dale separates Wenlock Edge from the southern hills where Brown Clee and Titterstone Clee rise from an undulating plateau at around 230m. Brown Clee (540m) is the highest point in Shropshire, and Titterstone Clee is only 7m lower.

Although the hills are the major landscape feature in west and south Shropshire, the dales and valleys that separate them, especially Corve Dale, are often wide and fertile. They are fed by countless streams which frequently cut steep narrow valleys of their own.

The northern and eastern plains are basically New Red Sandstone. Extensive glacial drift deposits cover much of the northern area, but though the associated hollows did give rise to meres, peat mosses and marshland, most have now been drained. The gently rolling landscape at around 80m is also broken by a line of higher sandstone hills — Nesscliffe, Pim Hill, Grinshill and Hawkstone — and in some areas sandy soils come to the surface. Moving south-east, the plain is slightly higher

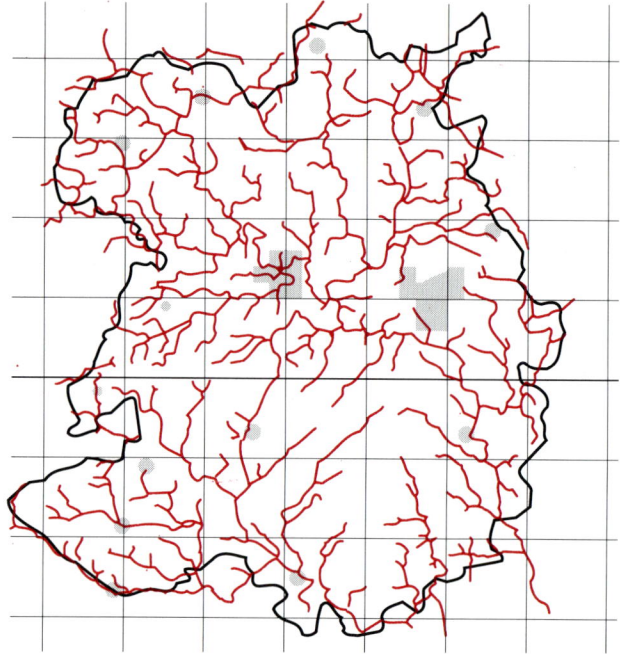

Map 2. Rivers, streams and canals. (© John Tucker 1988)

at about 120m, and here the countryside is more steeply rolling with deeper river valleys and sandstone outcrops.

The River Severn flows west to east then south-east across the southern parts of the plain, while its major tributaries from the north, the rivers Perry, Roden, Tern, and Worfe, drain the plain itself.

The rivers and main streams are shown in Map 2.

Relief

These natural foundations are reflected in the variation in the height of the land above sea level, shown in Map 3. However, the steep-sided hills and deep valleys create very considerable height differences over short distances. The Atlas is based on survey areas of 2-km squares, known as tetrads (see Appendix 1), and within the area of a single tetrad it is not uncommon to find a height difference of more than 150m, as shown in the two altitude maps (Maps 4 and 5). Several tetrads appear in both maps, as they contain both land above 427m and below 305m, or above 305m and below 183m. Some contain height differences of over 200m, and the maximum appears to be 290m where the whole western slope of Caer Caradoc rises from the Stretton valley to the summit at 459m.

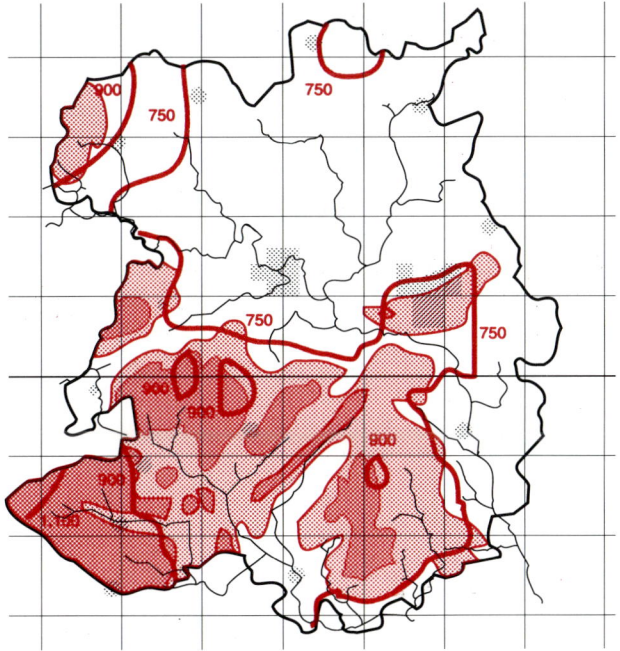

Map 3. Relief and rainfall.
(reproduced from *MAFF in the Midlands & Western Region,* 1988, © Crown Copyright, published by the Ministry of Agriculture, Fisheries & Food)

- more than 244 metres
- 122 to 244 metres
- 0 to 122 metres
- isohyets (mm)

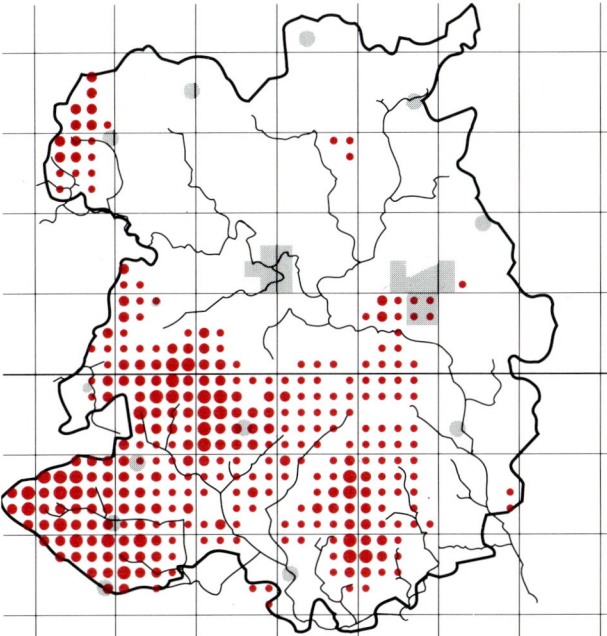

Map 4. Altitude.
- ● some land over 427m (1400 ft)
- ● some land over 305m (1000 ft)
- • some land over 183m (600 ft)

(adapted from *The Ecological Flora of the Shropshire Region,* Sinker et al, 1985; © Shropshire Wildlife Trust)

3

The predominance of the hills and plain ensure there is little low-lying land, and most of that below 61m shown in Map 5 is part of the river valley system of the Severn and its tributaries, the only major exception being the area to the east of the Tern valley, the ancient marshland of The Weald Moors.

Climate

The climate is intermediate between that of the warm south-east of England, the cooler, wetter north-west and the harsh condititions of the Welsh hills. There is considerable variation within Shropshire itself. While the uplands have heavy rainfall, much of the plain is in their rain-shadow (Map 3).

Being well away from the warming influence of the sea in winter, and its cooling influence in summer, winters are among the coldest in England, springs are late, and summers are relatively warm.

The agricultural growing season is therefore short and winters in the uplands of the south-west are especially harsh. Even the lower areas are well above sea level and when snow does fall it lingers.

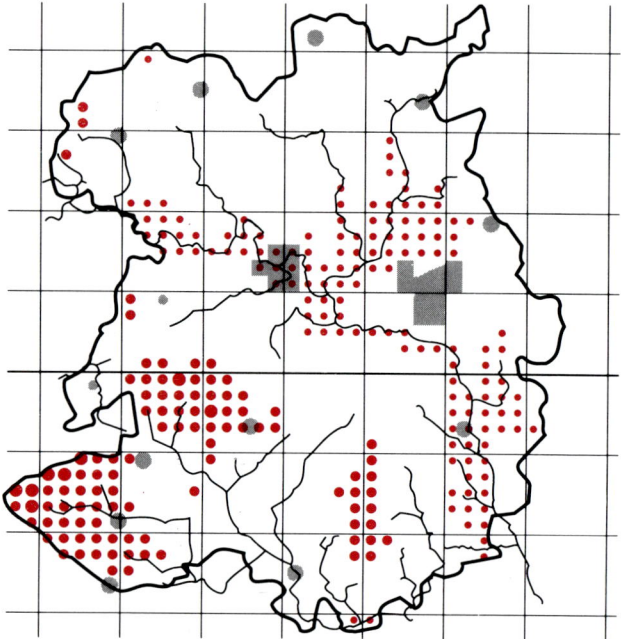

Map 5. Altitude.
- no land under 305m (1000 ft)
- no land under 183m (600 ft)
- some land under 61m (200 ft)

(ibid. © Shropshire Wildlife Trust)

LAND USE

The wide range of rocks and drift deposits has produced an equally wide range of minerals and soil types, which, coupled with the relief, climate, rainfall and drainage, has determined the land use.

Agriculture

Farming dominates the landscape, using 84% of the land. The national agricultural land classification "grades land according to the physical suitability of the area in which it lies for agricultural production" (MAFF 1988). The classification in Shropshire is shown in Map 6. There is no Grade 1 land. Grade 2 land occurs where there are well-drained soils, and no climatic and few relief considerations affecting agricultural use. Grade 3 land is widespread below 183m, and Grade 4 land has severe limitations for agriculture — it is generally above 183m with annual rainfall above 1270mm, or on the poorly drained soils of the coalfields and river valleys. Grade 5 is generally on the highest hills over 305m.

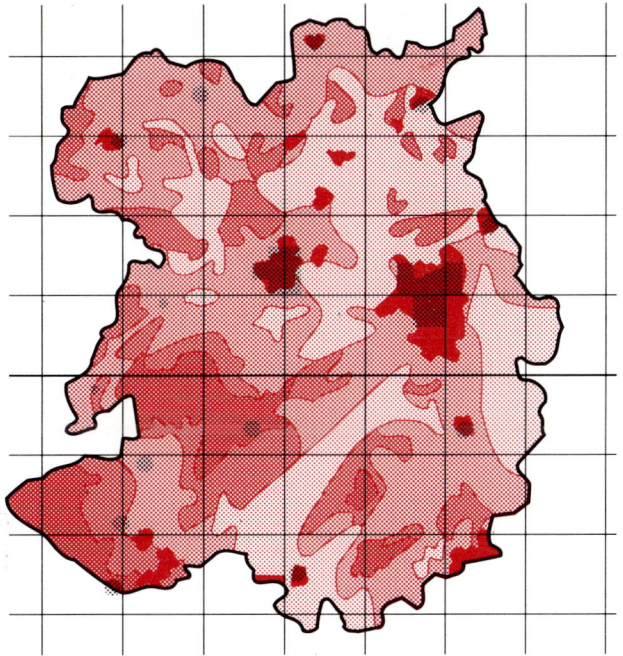

Map 6. Agricultural Land Classification.

- ▨ Grade 2 land
- ▨ Grade 3 land
- ▨ Grade 4 land
- ▨ Grade 5 land

(MAFF 1988 © Crown Copyright)

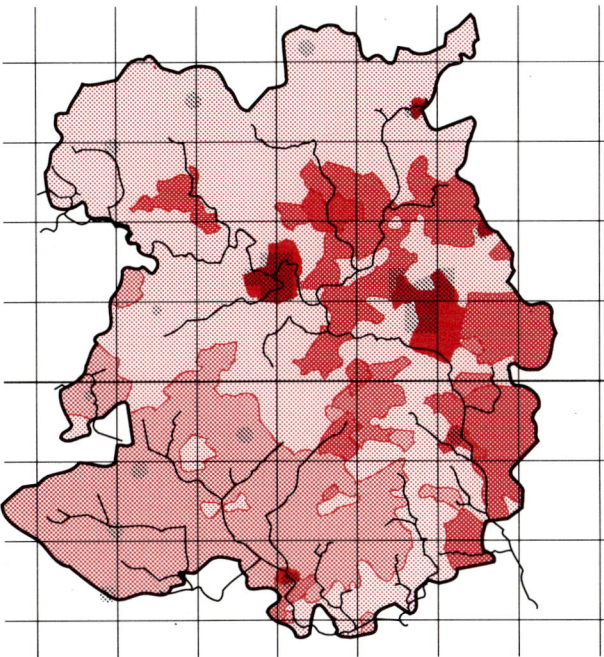

Map 7. Distribution of Predominant Farm Types.

- ▨ Dairy farming
- ▨ Livestock — mainly cattle
- ▨ Livestock — mainly sheep
- ▨ Cereals
- ■ Pigs and poultry
- ■ Urban areas

(MAFF 1988 © Crown Copyright)

The quality of land determines the type of farming, which in turn determines much of the landscape. The predominant farm types are shown in Map 7. However, the cultivated land on the mixed farms in the fertile valleys of the south and west, and scattered across the northern plain, does not show through on the map as it is swamped by the predominant livestock farming. Also, since the information for this map was collected in 1986, there has been a limited move away from dairy farming resulting from the introduction of milk quotas by the European Community. Land in use for cereal production is therefore under-represented on this map.

Woodland

Shropshire is well wooded, with 7.2% of its area still woodland (Map 8). Much is ancient semi-natural wood-land, with major remnants in the Wyre Forest, the Ironbridge Gorge, and along Wenlock Edge. Smaller open woods remain on other hillsides and riverbanks where the steep-ness of the slope has hindered agricultural clearance. Large coniferous woodlands have also been planted, especially on the hillsides of the south

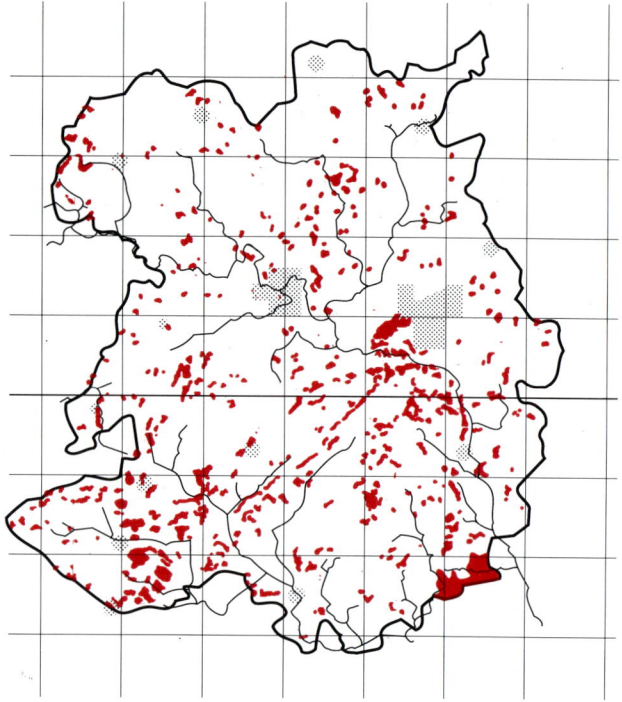

Map 8. Forestry and Woodland.
(Reproduced by courtesy of Shropshire County Council.
© SCC 1991)

and west, and small wood-lands, coppices and game coverts are still widespread.

Industrial influences

Apart from agriculture many other human activities have affected the landscape.

Mining used to be very important. West of the Stiperstones around Shelve used to be one of the richest sources of lead in Britain, and coal, iron, limestone, dhu-stone (basalt) and copper were all intensively mined in the Clee hills. New habitats have been created as a result — shafts, overgrown spoil tips, cliff faces and boulder-strewn slopes. More recently, here and elsewhere, quarrying for building and roadstone has created most of the cliff faces.

Similarly, interesting habitats have been created around Ironbridge and elsewhere in the Telford area as a result of industrial development based on the coalfield. The early iron industry's need to maintain a continuous supply of charcoal, through coppicing, is the main reason why the extensive ancient semi-natural woodlands of the Ironbridge Gorge have survived.

6

Apart from the natural meres, many open still-water habitats are also man-made. These include ornamental lakes around the old manor houses, pools resulting from gravel or sand extraction or subsidence of old mines, the construction of reservoirs and canals, and the extensive pools resulting from sugar beet processing at the Allscott factory. Ponds, drainage ditches and small reservoirs for livestock are a frequent by-product of farming, and trout fisheries are increasing.

Only Telford (population 107,700) and Shrewsbury (60,900) can be considered large towns. Both are predominantly suburban, with gardens and parks providing breeding habitats for the more common birds. Telford Development Corporation, charged with reclaiming the derelict land from the coalfield, aimed to invest sufficient funds in landscaping and new habitats to create an abundance of wildlife, though some of the "derelict" sites were preferable from the wildlife point of view.

BIRD HABITATS

The interaction of the landscape and land use has created a wide range of bird habitats.

Open moorland and upland heaths

Large areas of the highest tops in the Shropshire uplands and southern hills are open moorland, corresponding to the Grade 5 agricultural land (Map 6) and the highest hills (Map 3). They are mainly heathland dominated by heather, bracken and bilberry, with a few scattered stunted trees. On The Long Mynd the heather is cut or burnt periodically to promote new growth to benefit both sheep and Red Grouse, and some areas have been cleared to provide improved grassland. Small pools and bogs are frequent on the flat tops. There are some rock outcrops, with extensive boulder-fields on the Stiperstones and Titterstone Clee. As the hillsides are cut by fast-flowing streams in steep-sided valleys, single trees and small open woods can grow in very sheltered positions close to the summits. Gorse and other shrubs are also prevalent in some of these upland valleys.

Several different upland habitats and the birds associated with them exist side by side in a small area: Meadow Pipit, Skylark, Curlew and perhaps Red Grouse on the heather moor, Reed Bunting and Snipe in the wet flushes, Wheatear amongst the rocks and rabbit burrows, Whinchat and the occasional Stonechat in the bracken and gorse, Dipper and Grey Wagtail in the stream beds, and Tree Pipit, Redstart, Ring Ouzel, Pied Flycatcher, Wood Warbler, Buzzard and Raven amongst the scattered trees and woods.

The association of some of these 18 species is shown in Maps 9 and 10. Map 9 picks out those that inhabit the open moorland, Map 10 shows the upland woodland and stream birds, and the extent to which they exist side by side is indicated by the overlap between the two maps.

Upland grassland

Elsewhere on the high hills in the south and west, land has been improved for agriculture, mainly sheep grazing, and most is well drained with short well-cropped grass. Rock outcrops and rabbit burrows provide homes for the occasional Wheatear, but in general there are few ground-nesting birds. Apart from small

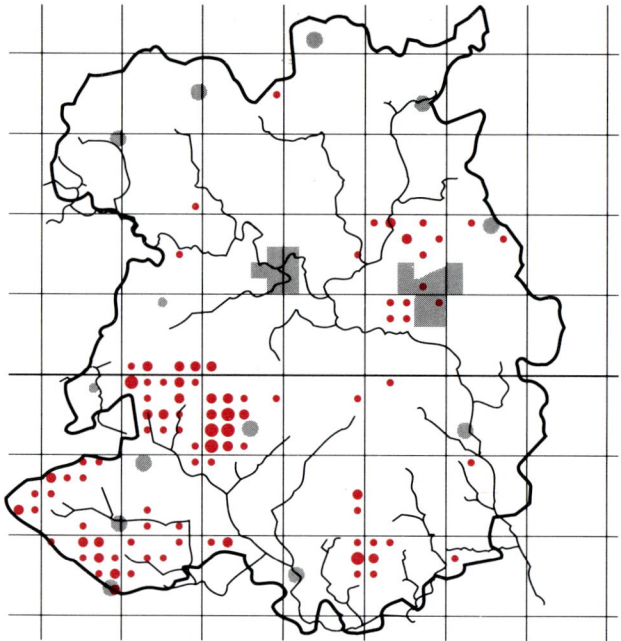

Map 9. Birds of Open Moorland.

Red Grouse, Snipe, Meadow Pipit, Skylark, Whinchat, Stonechat, Wheatear, Ring Ouzel and Raven: Probable or confirmed breeding.

- ● 8 or 9 species
- ● 5 to 7 species
- • 3 or 4 species

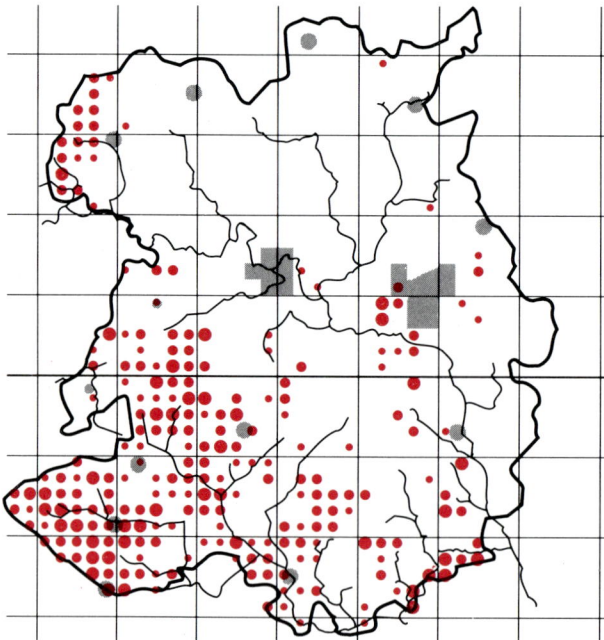

Map 10. Birds of Upland Valleys.

Buzzard, Tree Pipit, Grey Wagtail, Dipper, Redstart, Pied Flycatcher and Wood Warbler: Probable or confirmed breeding.

- ● 6 or 7 species
- ● 4 or 5 species
- • 3 species

numbers of Meadow Pipit and Skylark, birds inhabit such trees and unimproved land as may remain.

Upland field boundaries are usually fences rather than walls, providing little cover for nesting birds, but traditionally laid hawthorn hedgerows enclose many fields on the lower hillsides, providing numerous sites for hole-nesting birds, especially Redstart. The hollows between the hills are sometimes marshy, and the drainage ditches overgrown, so patches of rushes or longer grass may give some cover for Reed Bunting, Lapwing and the occasional Snipe.

The banks of the main streams and rivers draining the uplands are usually lined with trees rich in woodland birds, and mature hedgerows often enclose the fields alongside these streams. Most of the species found in tetrads dominated by upland grassland occur almost entirely in these habitats, which occupy only small areas along the valley bottoms. Many of the typical species are included in Map 10.

Lowland grass and pasture
Descending the northern slopes of the uplands onto the plain and down into the Severn valley around Shrewsbury, the land is richer and used mainly for dairy farming, though increasingly cereals are grown. Lapwing, Curlew and Skylark nest here. This land use continues northwards, and then extends eastwards to the north of the line of sandstone hills (Maps 1 and 7).

Many scattered small woods, coppices and game coverts, and large numbers of mature hedgerow trees, provide cover for many woodland and scrub birds.

Cultivated farmland
Most of east and north-east Shropshire is heavily cultivated. The main crops are barley and wheat, but these are usually rotated with potatoes, cabbage, peas and especially sugar beet. Yellow Wagtail and Corn Bunting nest on the land here, and Map 11 shows the tetrads where both species breed. This distribution corresponds closely with the areas given over to extensive cereal farming (Map 7), though other sites are provided by cultivated land on the scattered mixed farms elsewhere.

Though there is a tendency to remove hedgerows to create larger fields, Shropshire has suffered less in this respect than many counties. Fields are still often enclosed by mature hedges, and small woods and mature hedgerow trees also remain, so there is some diversity of bird life even in tetrads where the land is predominantly cultivated.

Woodland
The deciduous and mixed woods support a wealth of wildlife and have high numbers of breeding birds. However, the typical woodland species, such as Great Spotted Woodpecker, Nuthatch and Treecreeper, also use small coppices and scattered trees, and Map 12 showing their association demonstrates the widespread distribution of trees as well as the large areas of woodland (see Map 8).

Trees in the coniferous plantations are close together, so the mature woodland is dark and impenetrable, devoid of all except the most adaptable or catholic birds — Woodpigeon, Goldcrest, Coal Tit, Blue Tit and Chaffinch. However, when the trees are first planted, and again after thinning out, more open and interesting habitats are

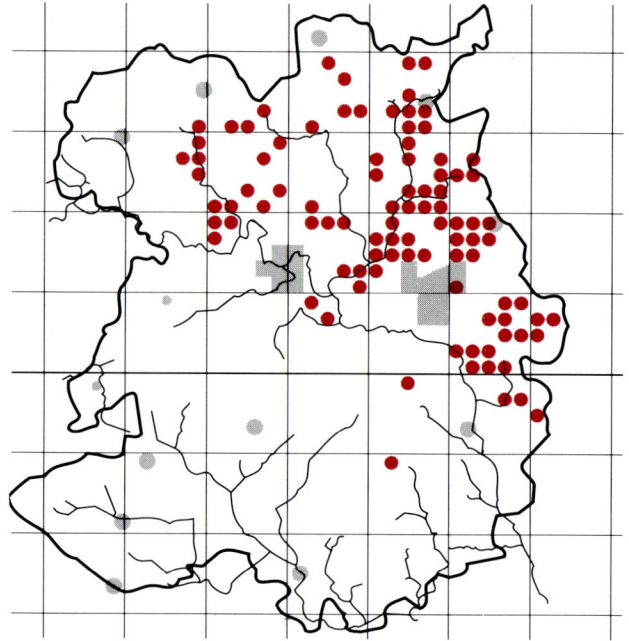

Map 11. Birds of Arable Farmland.
● Yellow Wagtail and Corn Bunting: Probable or confirmed breeding of both.

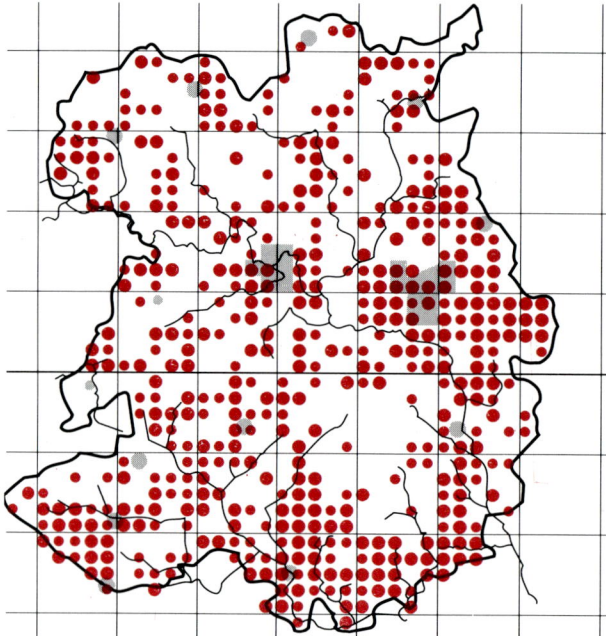

Map 12. Woodland Birds.
Great Spotted Woodpecker, Nuthatch and Treecreeper: Probable or confirmed breeding.

● all 3 species
● any 2 species

created, offering opportunities to Tree Pipit, Whinchat, Grasshopper Warbler and occasionally Redpoll.

Rivers and streams

The major rivers and tributaries are included on the background map for every species. They, and all significant streams, are shown in Map 2.

The Severn is far and away the largest river. On the western border its confluence with the River Vyrnwy has created an extensive lowland floodplain. This plain narrows until the river flows out to the east of Shrewsbury, after which it widens considerably at the confluence with the River Tern, and remains so as the river meanders to Buildwas. After the 15km of the steep-sided Ironbridge Gorge the valley again opens out at the confluence with the River Worfe. These river systems have created most of the land below 61m (Map 5), and large sections flood every winter, remaining damp into the breeding season.

The Severn and its main tributaries are now managed by the National Rivers Authority (and formerly by the Severn-Trent Water Authority) to improve the flow and minimise flooding. The Severn especially is prone to sudden changes in water level following heavy rain further upstream. At the beginning of the breeding season the banks seem rather bare but as the water level falls a rich margin of emergent and bank-side vegetation appears with large expanses of water crowfoot in the shallower riffles between the deeper reaches. On some stretches willow and alder add to the diversity, attracting Sedge Warbler and Reed Bunting, while Kingfisher and Sand Martin make nest holes in the steep banks.

In contrast, the rivers that rise in the hills and flow south of the watershed into the Clun/Onny/Corve/Teme system have been less intensively managed. They are generally boulder-strewn, faster flowing and shallower, with some relatively deep pools where the gradient temporarily slackens off. The banks are lined with trees and many stretches are overgrown and inaccessible. These rivers and streams are rich in bird life, and it is not uncommon for the same stretch of river to have both pools that support Kingfisher and rapids occupied by Dipper and Grey Wagtail.

Elsewhere, the same characteristics are shown by the streams flowing from the Oswestry uplands, those streams that descend steep valley-sides into the Severn around Ironbridge, and the upper reaches of the rivers Tern and Worfe. Map 13 show those tetrads with probable or confirmed breeding of both Kingfisher, and Dipper or Grey Wagtail.

Open water

The natural meres, mainly around Ellesmere, but also Fenemere and Marton Pool and stretching south to Berrington Pool south of Shrewsbury, are the largest areas of open water. These are supplemented by the network of canals in the north, man-made ornamental lakes around the widely scattered old manor houses and parks, such as Shavington, and balancing lakes and landscape features around Telford New Town. The largest reservoir, at Chelmarsh, is a reserve with access restricted to members of the SWT and SOS. A managed reed bed, introduced into the marsh at its northern end, is the only place where Water Rail is certainly known to breed regularly.

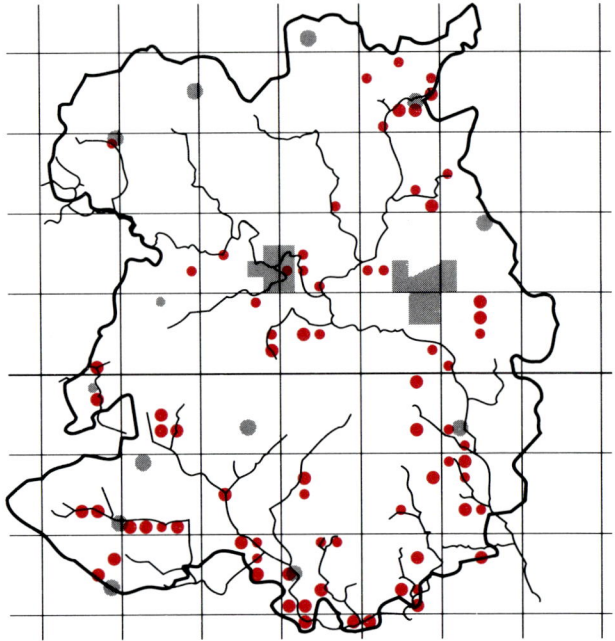

Map 13. Birds of Rivers and Streams.

Kingfisher, Grey Wagtail and Dipper: Probable or confirmed breeding.

● all 3 species
● Kingfisher and either Grey Wagtail or Dipper.

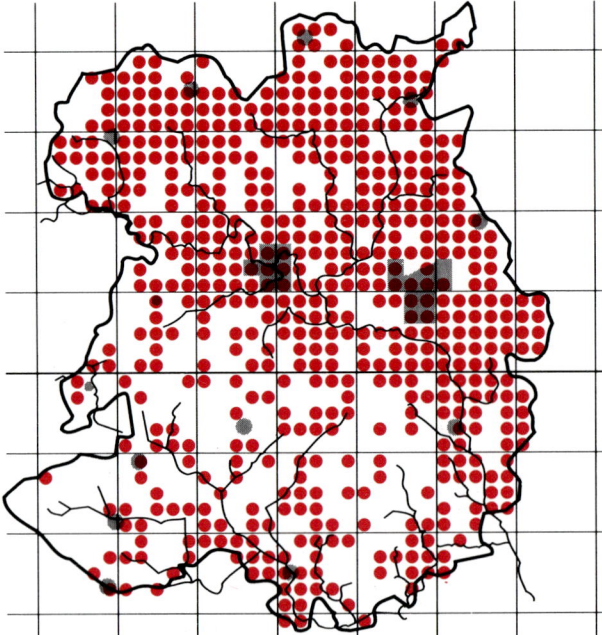

Map 14. Birds of Small Pools.

●Mallard and Moorhen: Probable or confirmed breeding of both.

12

Sand and gravel removal have also created several interesting pools and wetlands, not least adjacent to the SOS bird reserve at Venus Pool. Though Venus Pool itself was created by poor drainage, it was used as a settling pool for sand extraction, which added to the diversity of habitat, and the reserve now boasts a wide range of breeding waterfowl and wading birds, including some of the more unusual species — Oystercatcher, Little Ringed Plover and Redshank.

There are few large waters in the south and west, but of particular note are Shelve Pool at over 300m, where the surrounding reed beds support Reed and Sedge Warbler, and Reed Bunting, and Boyne Water at 455m near the top of Brown Clee, with breeding Moorhen and Tufted Duck. Coot also bred at Boyne Water in the earlier part of the Atlas period, though clearance of all the emergent vegetation has rendered it unsuitable now.

Small pools are widespread on the plain, and in the valleys between the uplands, as reflected in the large number of tetrads where both Mallard and Moorhen breed (Map 14).

The importance of open water as a breeding habitat for birds is dependent on extent and quality of both the water and the surrounding vegetation, and also on the

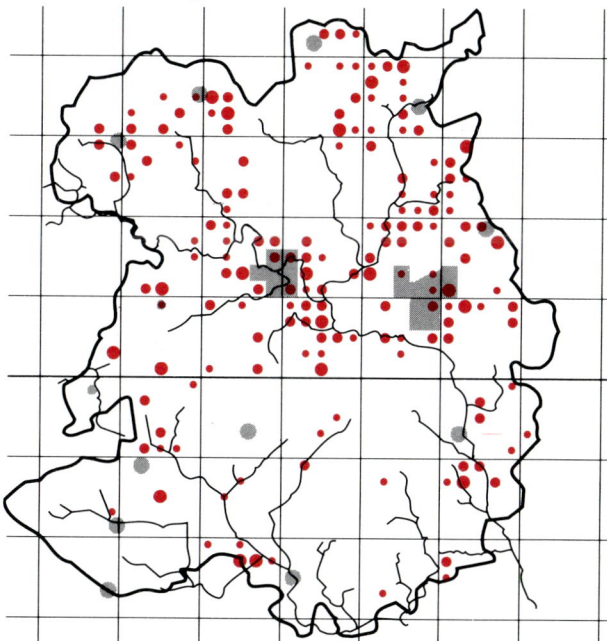

Map 15. Birds of Open Water.
Little Grebe, Great Crested Grebe, Mute Swan, Canada Goose, Tufted Duck and Coot: Probable or confirmed breeding.
● 6 species
● 4 or 5 species
• 3 species

steepness of the banks. The pools with the best range of breeding birds have good cover to the water's edge, gently sloping banks which give marshy areas with reeds or rushes, emergent vegetation and overhanging trees. The association of widespread species using an open-water habitat — Little Grebe, Great Crested Grebe, Mute Swan, Canada Goose, Tufted Duck and Coot — is shown in Map 15.

Lowland mosses and bogs

Originally the north Shropshire plain contained many meres and "mosses" — meres which have become fully overgrown through the natural encroachment of successional vegetation, forming peat. Rowley (1972) commented that "about four fifths of the peatland acreage of north Shropshire is now wholly reclaimed and under pasture" and more has been drained since. The only extensive remnants are Whixall Moss National Nature Reserve and Wem Moss SWT reserve, both on the border with Clwyd.

The Weald (i.e. Wild) Moors north of Telford are extensive mosses lying lower than the surrounding plain. Map 5 shows them as the large area below 61m to the east of the River Tern. They have proved difficult to drain completely and retain some of the original moss wetland character.

These three areas are the only major remaining lowland habitat for Whinchat, while the mosses also support both Tree and Meadow Pipit, and The Weald Moors have very small numbers of both Snipe and Redshank.

The scarcity of lowland wetland sites is illustrated by the association of Snipe, Redshank, Oystercatcher and Little Ringed Plover in Map 16. The last two also tend to occupy different habitats from the first two, and continuing drainage of wetlands is a major threat to Snipe and Redshank. Map 16 therefore emphasises tetrads where these two species both probably breed, and also shows where they both possibly breed. In some of these tetrads probable or confirmed breeding has been established for one of them, but not the other. Only one additional dot (at SJ50P, which has probable breeding of Little Ringed Plover as well as Redshank) would appear on the map if it showed probable breeding of any two of these species.

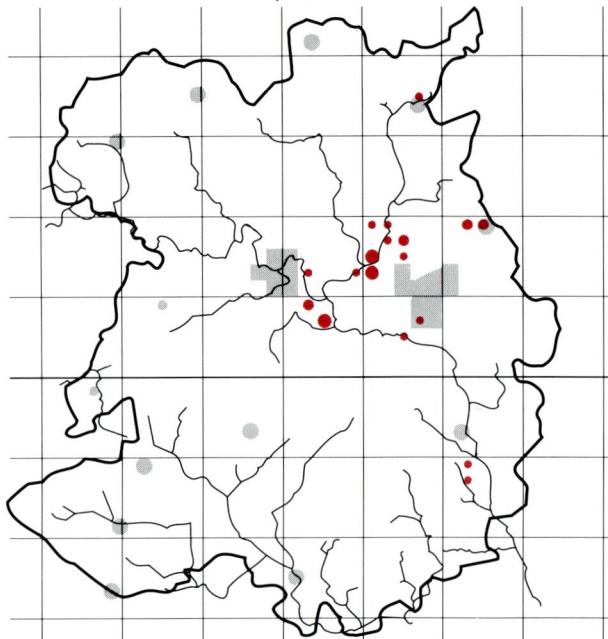

Map 16. Birds of Lowland Wetland.
Emphasising shortage of sites for Snipe and Redshank.
● Probable or confirmed breeding of Snipe, Redshank, Little Ringed Plover AND Oystercatcher
● Probable or confirmed breeding of both Snipe and Redshank
• Possible breeding of both Snipe and Redshank

Lowland heath

This once covered the larger part of the light sandy soils of the north Shropshire plain, but most has been cleared for agriculture and only a few relics, such as Prees Heath, Steel Heath, Hodnet Heath and parts of Haughmond Hill, remain.

The old industrial pit mounds from the intensive mining around what is now Telford frequently reverted to heathland once disturbance ceased, though it has largely been lost again in recent years. These areas provide much of the lowland habitat of Meadow Pipit.

THE CHANGING FACE OF SHROPSHIRE

Reference has already been made to the extent to which agricultural improvement has created the Shropshire landscape. Enclosure of the land, followed by drainage and other human influences, have been well documented (Rowley 1972).

These changes are still going on. *Losing Ground in Shropshire* (SWT 1989) describes the way in which over 5000 sites were surveyed and the best 766 identified as "Prime Sites for Nature Conservation" in 1978–79. By 1989, 95 (12%) had been destroyed, and 133 (17%) had been damaged. Of this destruction, 64% was due to agriculture, 21% to forestry, and 15% to development.

Agricultural change

MAFF (1988) has documented a strong trend to increased farm size between 1978 and 1986, coupled with increased purchase of farms by financial investment institutions. Only 65% of farmland in the west Midlands was owner-occupied in 1986. Investment in machinery has increased while the workforce has decreased, and larger fields and destruction of hedgerows have been an inevitable consequence. Farming is now often an industrial activity, rather than a rural one in sympathy with the landscape and wildlife.

Hopefully the anticipated reduction in farm subsidies, and expanding schemes such as the Environmentally Sensitive Areas programme and Countryside Stewardship, will reverse this trend, though they have come much too late to save many of the best habitats.

Apart from the destruction of habitat, changing farming practices are also affecting bird populations in other ways. The change from spring to autumn planting of wheat and barley has substantially reduced the area of stubble that sustains several species of seed-eating birds through the winter, and also the bare ground that Lapwings in particular need at the start of the breeding season. In addition, the earlier cutting of hay and increasing use of silage destroy many nests each year and have almost totally removed potential Corncrake habitat. These changes are affecting other species as well, while use of herbicides and pesticides seriously affects many more, either through eliminating their seed food or insect prey, or poisoning the birds themselves as chemicals accumulate in their body tissue. There is now overwhelming evidence that the breeding numbers of many farmland species are declining rapidly, and the scientific research which produced it is fully documented in *Population Trends in British Breeding Birds* (Marchant *et al.*) published by the British Trust for Ornithology (BTO) in 1990.

Ponds, another important habitat, are also being lost as they are infilled, largely for agricultural intensification. For example around Shrewsbury there were 490 ponds at the turn of the century, 360 in 1963, and only 50 by 1990 (Jepson 1991).

Urbanisation

Farming is not the only cause of habitat loss. The population of the area covered by Telford New Town was intended to double following the creation of the Development Corporation in the 1960s, and many of the SWT Prime Sites lost in the last ten years were in the New Town. They totalled about 10% of the area lost, with woodland, heath and flower-rich grassland/tall herb categories suffering particularly. Other towns and villages are continuing to expand, and while careful planning may increase some habitats, such as gardens, parkland and ponds, this will mainly benefit the more common birds that adapt to suburban life, and will not help the less common and more specialised species. Even here, the increasing use of herbicides and pesticides by gardeners is likely to suppress bird populations.

Growing suburbanisation was also indirectly responsible for damage to Whixall and Fenn's Mosses, where the peat was being cut for sale in garden centres. Fortunately, total destruction of this site has been averted, as a strong local campaign led to purchase and management as a National Nature Reserve by English Nature, but this is an example of the kind of threat that remains elsewhere.

New roads to cater for the increasing levels of traffic destroy much valuable habitat, though in some cases the large embankment verges are better and provide more variety than the agricultural land lost. The new Shrewsbury by-pass cuts through woodlands very close to a heronry and it remains to be seen if the Herons will be able to cope with increased disturbance. Many birds are killed by traffic collisions, and, as the species account demonstrates, the Barn Owl suffers substantial losses from this cause, especially along the Oswestry by-pass.

Leisure activities
Many species are vulnerable to disturbance from increased human use of their living quarters. If scared from their nest, some will abandon it altogether, while others may return later, only to find the eggs or nestlings were predated in the meantime. Nests at the water's edge are prone to flooding, or damage by the wash from boats. If eggs or young are deserted or lost too late in the season for the adults to lay again, reproduction is lost for that year, and the population will suffer.

Cars allow an ever increasing number of people to enjoy the countryside and a whole range of activities — fishing, rambling, climbing, orienteering, riding, pony trekking, mountain biking, motocross, school field trips, canoeing, water-skiing, raft-racing, and even birdwatching — all increase disturbance of nesting birds.

New development for leisure activities also threatens important sites. Prees Heath, one of the few remaining heathlands, is now at risk from a golf course and associated plans, as well as proposed gravel extraction and further ploughing, and another local campaign has been launched to save it.

WILDLIFE PROTECTION
The changing face of Shropshire, and changes in human activity, are obviously affecting bird habitats and populations, and the consequences will have to be carefully monitored and assessed.

The Wildlife and Countryside Act (1981) provides legal protection for most wild birds and their nests, but enforcement is difficult; MAFF and the RSPB have reported a considerable number of cases of illegal use of poisons and traps within Shropshire. Unfortunately the law does little to protect bird habitats unless they are part of a nature reserve. Even nationally important sites that have been given statutory protection by designation as Sites of Special Scientific Interest (SSSI) have been damaged here. The Prime Sites recorded by the SWT have no legal protection at all, though they do receive some attention in the County Structure Plan, and increasingly in Local Plans, and their survival depends almost entirely on the attitude of the landowner.

More important is the fact that most birds, even the less common ones, nest outside the Prime Sites and SSSIs, and once again the attitude of the landowner is all-important. For example, hedgerow and scrub clearance during the breeding season can destroy large numbers of nests, but it is usually possible to persuade owners to wait until the young have flown. It is much harder to persuade people that similar actions at other times of the year, or draining or ploughing a meadow, are just as deadly in the long run because the parents will be homeless when looking for a nest site the following year. Many landowners are making real efforts to improve the habitats on their land by more sympathetic management of hedgerows, planting up headlands and establishing pools, but it remains a fact that the protection of most of our bird populations is in the hands of individual landowners and managers.

It is in this context that the SOS, along with the SWT, will continue to monitor both the population level and distribution of our birds. We will assess the effects of the changes described above, particularly those of land management upon habitats, and, wherever necessary, seek to influence policy both locally and nationally, and take action to control the worst excesses.

THE SHROPSHIRE BREEDING BIRD ATLAS

The primary purpose of the Shropshire Breeding Bird Atlas is thus to provide a bench-mark against which future trends in distribution and population can be monitored. The commentary with each species map relates distribution to the various habitats described above, estimates the current population levels, and wherever possible compares the current position with previous assessments. The results must be interpreted with care however, and the limitations of this type of survey are described more fully in Appendix 1.

Even so, judging from the results, it is clear that some species are increasing, many more are declining, and a few have disappeared as breeding birds in Shropshire during the last quarter-century, as shown in Tables 1, 3 and 4 in the conclusions to the species accounts. The current status and protection given to Shropshire's breeding birds is shown in Appendix 2. Although some species are known to be increasing, their growth in numbers is no compensation for the decline or loss of others.

The SOS aims to "encourage the study and protection of birds of Shropshire". The Atlas has provided invaluable base data for this task. Work is now advancing to ensure that all the best sites found by Atlas fieldwork are included in the SWT's Prime Sites system.

By building upon the records presented in this Atlas, farmers, landowners, planners, conservationists, ornithologists and amateur bird watchers can all help to safeguard the future for birds in Shropshire.   *LS*

FURTHER READING:
1. *The Making of the Shropshire Landscape*   Rowley 1972
2. *MAFF in the Midlands & Western Region*   MAFF 1988.
3. *The Geography of Shropshire*   Sinker in *A Handlist of the Birds of Shropshire*
   SOS 1964.
4. *Shropshire's Wild Places*   Jenkinson 1992 (in prep.).
5. *Where to Watch Birds in the West Midlands*   Harrison and Sankey 1988.
6. *Wildlife in Telford*   Telford Development Corporation 1981.
7. *Losing Ground in Shropshire*   SWT 1989.
8. *Population Trends in British Breeding Birds*   BTO 1990.

THE SPECIES ACCOUNTS

The species accounts illustrate and describe the distribution of Shropshire's breeding birds.

Maps

The maps are based on detailed surveys over six years from 1985 to 1990. They show only what has been found by Atlas fieldworkers, or reported to the County Bird Recorder during this period.

For comparison, Map 17 shows the main rivers and towns, and the Ordnance Survey (OS) National Grid 10-kilometre squares. These are included as background to all the species maps so distribution can be related to local landmarks and OS maps.

Each dot on a species map means it was recorded somewhere in the 2-km square, known as a tetrad, at that point on the map. The large dots represent confirmed breeding, the medium dots probable breeding, and the small dots possible breeding. The definitions of confirmed, probable and possible breeding, the fieldwork methods used and the way the maps have been compiled are all described in Appendix 1.

Maps are not published for seven rare species which are vulnerable to disturbance or illegal persecution, and whose breeding sites must be kept

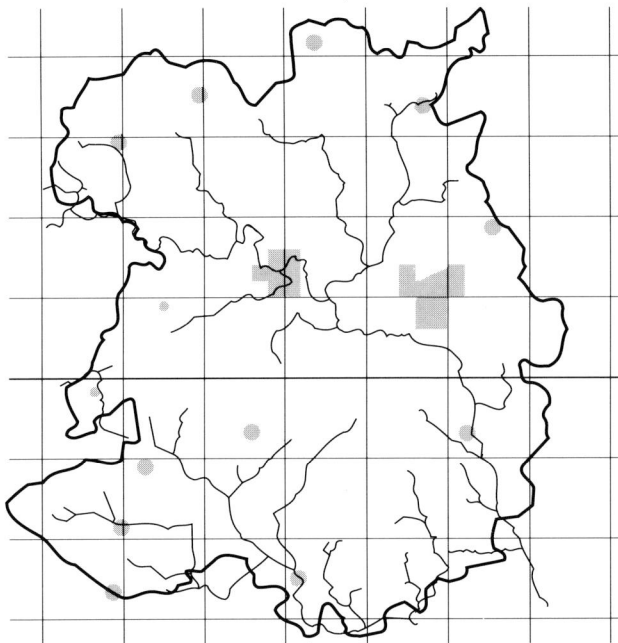

Map 17. The background map to the species accounts, showing the main rivers and towns.

secret — Black-necked Grebe, Marsh Harrier, Hen Harrier, Goshawk, Merlin, Hobby and Peregrine. Neither are maps published for Corncrake, which no longer breeds in Shropshire, nor for several species that were recorded on Atlas cards but which are only winter visitors or passage migrants. Anyone chancing across breeding sites for any of these species, or any others not included in the Atlas, should report the information in confidence to the County Bird Recorder, but otherwise keep it to themselves.

Breeding season status

The status of each species, but only in the breeding season, is given under its title. No account is taken of other times of the year, though if appropriate this may be referred to in the text. Species recorded as probable or confirmed breeding in every year of the Atlas period are described in the heading only as either "resident" or "summer visitors". Resident means that at least a part of the breeding population is found in Shropshire for at least some of the winter months. In most cases, resident birds are sedentary, and there is no difficulty in defining their status, but, for example, Meadow Pipit and Goldfinch are described as resident, even though a high proportion of them go to Iberia for the winter, and Lapwing may disappear altogether in really hard winters.

No judgement is made here on abundance — for example, the Hobby was found throughout the period at only one site, but it is described in precisely the same way as the Swallow, as a summer visitor.

Species for which probable or confirmed breeding was not established every year are given the breeding status "regular" if this was established in at least 4 years, "occasional" if in 2 or 3 years, and "rare" if in only 1 year, providing breeding was confirmed at least once.

The only exceptions to this are a few scarce and elusive species — Water Rail, Long-eared Owl, Redpoll and Hawfinch — that are likely to be overlooked and which are believed to have a small breeding population every year, and which are described as resident.

Where breeding was not confirmed at all, the breeding status is described as possible or probable, according to the best evidence obtained by fieldwork.

One species expanding its range — Peregrine — became established during the Atlas period. Though now almost certainly a resident, it did not breed here in the first two years, and is therefore still described as regular rather than resident. Similarly Mandarin and Siskin, and perhaps Barnacle Goose, Gadwall, Shoveler and Goosander are variously described as occasional or regular, though they may now be resident. Conversely, a breeding population of the resident Shelduck may not have been present every year, as some of the records may relate to pairs that had not reached maturity.

Distribution statistics

The statistics show the number of tetrads in which each level of evidence — possible, probable or confirmed breeding — has been established, and the total. The percentages show those numbers as a proportion of all the 870 tetrads in Shropshire, indicating how widespread the species is. It would be necessary for a species to be recorded in every tetrad for the percentages to add up to 100%. If it was recorded at some level in half the tetrads in the county — a total of 435 — then the percentages would add up to 50%.

Comparing the proportions gives an indication of the relative difficulty of proving breeding. As each figure, including the total, is calculated as a percentage of 870, then rounded to the nearest whole number, some of the total percentages shown are slightly different from the sum of the three levels of breeding evidence.

Beware making comparisons between these statistics and those included in atlases for other counties, as some of them express the number of tetrads in which each level of breeding was established as a proportion of the total tetrads in which the species was recorded — i.e. the percentages add up to 100% by definition even for scarce species recorded in few tetrads.

The number of possibly breeding records is omitted in the case of some scarce raptors, for which no map is published. As they hunt over large distances, the figure would be misleading.

Commentary

The accompanying text is specifically about the position in Shropshire, and comments may or may not apply elsewhere, though the context should make this clear. It explains the distribution, interprets the map, and outlines the status and breeding habitat of each species. Where possible this is related to the habitat maps. The convention is used of giving a capital letter to each word in any place name shown on the OS map, for example "The" is part of the name of The Long Mynd, The Wrekin and The Weald Moors but not the Stiperstones; and the north Shropshire plain and Clee hills are general areas, not specific locations named on the map. All place names and geographical features mentioned are listed in Appendix 5, along with the 4-figure OS map grid reference that identifies the 1-km square, so all locations can be related to the distribution map if required.

The full scientific name of any plants or animals referred to are listed in Appendix 4. Descriptions of the birds themselves, their behaviour and how to identify them are omitted as they can be found in any of several excellent field guides.

The accounts draw heavily on two publications by the British Trust for Ornithology (BTO) — *The Atlas of Breeding Birds in Britain and Ireland* (Sharrock 1976), based on fieldwork from 1968–72, and *Population Trends in British Breeding Birds* (Marchant *et al.* 1990). Accounts for the rarer species draw on *Red Data Birds in Britain* (Batten *et al.* 1990), published for the Nature Conservancy Council and the Royal Society for the Protection of Birds (RSPB) in 1990. These publications are referred to in the accounts as "BTO *Atlas* 1976", "*Pop. Trends* 1990" and "*Red Data* 1990" respectively. They include extensive references to original research and, in the main, only references not listed there are included in the species accounts in this Shropshire Atlas. Anyone wishing to study any species further, or verify the statements made, should refer to these publications for the relevant source. All reference given in the accounts are listed in full in Appendix 3. Comments attributed to these publications generally apply to Britain as a whole and not specifically to Shropshire.

The distribution of every species has been compared with that described in *A Handlist of the Birds of Shropshire* (Rutter *et al.* 1964), published by SOS, based on records up to 1963, and referred to as "*Handlist* (1964)"; the Shropshire part of the map in the BTO *Atlas* (1976); and the annual *Shropshire Bird Reports* ("*SBRs*") between 1972 and 1985. Any apparent change in the range or population since then is explicitly stated. If no such statement is made, there is no significant evidence of any change in status at the present time, though of course this may be due to lack of research or historical data, rather than stability in the distribution.

The BTO *Atlas* (1976) was based on surveys of 10-km squares, rather than the more thorough coverage of this Atlas. Where the species accounts refer to 10-km squares they are comparing current results with those from this earlier national *Atlas*. The relationship between tetrads and 10-km squares is described in Appendix 1.

Population estimate

An estimate of the breeding population is given, but it must be stressed that no attempt was made to establish this during the fieldwork. In many cases the number of breeding pairs per tetrad has been estimated, based on information in *Pop. Trends* (1990), or from local results from national BTO surveys, or impressions gained from Atlas fieldwork. This has been multiplied by the number of tetrads in which the species was recorded at the appropriate level. For species that occur commonly throughout the county, this will be all 870 tetrads, even if there are a few gaps on the map. For many species only "occupied" tetrads — defined as meaning tetrads with probable or confirmed breeding — are taken into account, particularly if territories are easy to locate but breeding is difficult to confirm, and possible breeding records might refer to migrants. For other more elusive species all records are likely to relate to breeding pairs, while scarce breeding birds that migrate through the county in larger numbers will be over-represented. The estimate may also have been adjusted to take into account whether the particular species is likely to be under- or over-recorded, for example because it is very elusive or nocturnal, or each pair has a large territory likely to cover several tetrads.

In some cases only the most crude of estimates has been possible. The most recent British population estimate, and any subsequent change indicated in *Pop. Trends* (1990), has been applied to Shropshire in direct proportion to the respective areas occupied, usually based on the distribution map in the BTO *Atlas* (1976), and perhaps with an adjustment where the county provides clearly better or worse habitat than the national average. The national estimates themselves and their application to Shropshire are likely to contain large margins of error.

The assumptions made in each calculation are described in the account, so future research can verify or amend the population estimate. Nevertheless, though in many cases they are "guesstimates", estimates now given here are believed to be broadly correct. Unless stated, no previous attempt has been made to establish the county population.

Abbreviations used

The *Handlist* (1964), BTO *Atlas* (1976), *Pop. Trends* (1990), *SBRs* and *Red Data* (1990) are publications described above. The other abbreviations used are as follows:

Winter Atlas	*An Atlas of the Wintering Birds of Britain and Ireland* (Lack 1986), published by BTO.
BWP	*The Birds of the Western Palearctic* (Cramp & Simmons), published by Oxford University Press.
CBC	Common Bird Census, an annual survey organised by the BTO to monitor the population level of many species. *Pop. Trends* (1990) is largely based on the results of the CBC since 1964.
The Flora	*The Ecological Flora of the Shropshire Region* (Sinker *et al.* 1985) published by the Shropshire Trust for Nature Conservation (now Shropshire Wildlife Trust), which includes a tetrad atlas of the county's plants.
MAFF	The Ministry of Agriculture, Fisheries & Food.

RSPB	The Royal Society for the Protection of Birds.
RSPCA	The Royal Society for the Prevention of Cruelty to Animals.
SOS	The Shropshire Ornithological Society.
SWT	The Shropshire Wildlife Trust, formerly STNC, the Shropshire Trust for Nature Conservation.
WBS	Waterways Bird Survey, an annual survey of water and waterside birds organised by the BTO, similar to the CBC.

Authors

Each commentary is attributed to its author by initials at the end. Accounts have been written by Tony Cross (AVC), Allan Dawes (APD), John Milner (JM), Jack Sankey (JS), Leo Smith (LS), Derek Sparkes (DS), Gerry Thomas (GT), Tom Wall (TW), Michael Wallace (MW) and Colin Wright (CEW).

Summary

Some gaps in the distribution maps are inevitable in covering such a large county with a relatively low human population. Several species breed all over the county, and there appears to be suitable habitat in every tetrad. The gaps that appear in the maps for these species — Woodpigeon, Wren, Dunnock, Robin, Blackbird, Blue Tit, Magpie, Carrion Crow and Chaffinch — are believed to be due to absence of observers at the right time, rather than the absence of the birds themselves. Most other species will be similarly under-recorded to some extent.

Some of the maps must be interpreted with caution. They are based on actual fieldwork results, and other factors may also influence them, as described in Appendix 1. However, the maps and species accounts together give an excellent indication of the current distribution of Shropshire's breeding birds.

LITTLE GREBE
Tachybaptus ruficollis

Status: **Resident**

Tetrads with evidence of breeding

Confirmed	45 (5%)
Probable	38 (4%)
Possible	38 (4%)
Total	**121 (14%)**

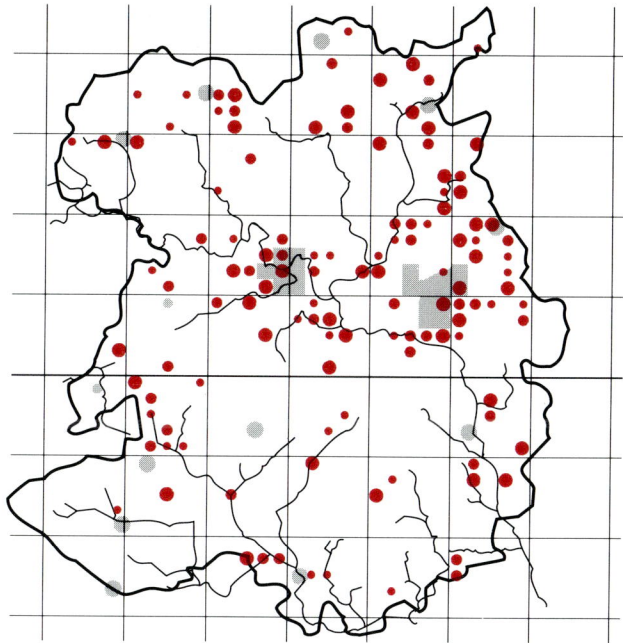

The "whinnying" courtship display call of Little Grebes is often the first indication of their presence, especially where shallow water is thick with vegetation. Resident on lakes, pools and canals, they will use much smaller waters than the Great Crested Grebe, which explains the wider scatter of occupied tetrads when compared with that species. Although they appear on the River Severn during the winter months, few remain there to breed.

Little Grebes are easily overlooked, and confirmation of breeding can be difficult as adults and chicks spend a considerable amount of time hidden in the emergent vegetation. Apparent absence from some tetrads with suitable habitat deserves further investigation, although at some sites competition with the larger Great Crested Grebe may be a reason. Some upland pools are occupied, like Shelve at 300m, and breeding was confirmed on Stapeley Hill at 310m and Brown Clee at 280m. No records exist for the higher pools on The Long Mynd which, in common with many pools in the hills, have inadequate emergent vegetation.

There is little evidence of change in national population levels at the present time and hard winters are probably the main controlling factor (*Pop. Trends* 1990). Varying water levels can affect breeding success, especially at sites like Brown Moss where, in hot summers, some of the smaller pools dry out completely.

The larger waters can support several pairs whilst the smaller pools may hold just one. Assuming each occupied tetrad holds an average of three to four pairs gives a population of between 250 and 330 pairs.

CEW

GREAT CRESTED GREBE
Podiceps cristatus

Status: **Resident**

Tetrads with evidence of breeding

Confirmed	52 (	6%)
Probable	19 (	2%)
Possible	17 (	2%)
Total	**88 (10%)**	

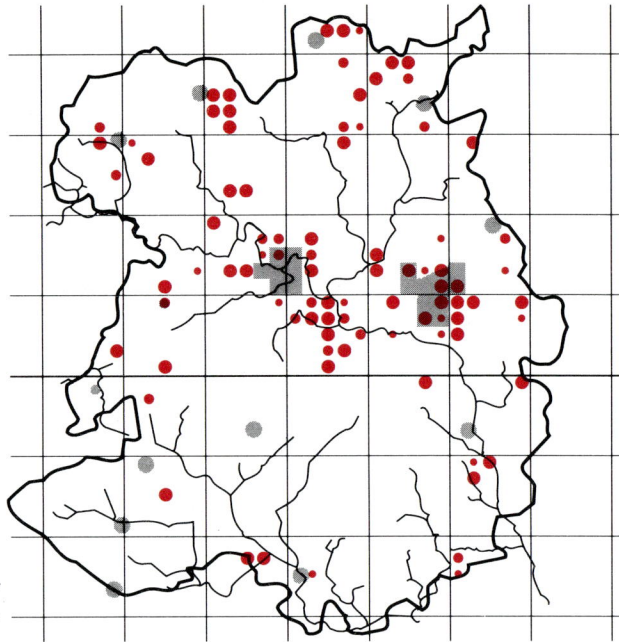

The Great Crested Grebe is resident on the larger lakes and meres. Slow reaches of the Severn occasionally hold pairs but the variable river levels, especially in wet years, make nesting hazardous as nests are often flooded and washed away. The courtship display is a joy to watch and the loud calls make pairs easy to locate. Some pairs raise two broods and the stripy young are obvious as they beg for food or ride on the adult's back, so confirming breeding is not difficult.

Human persecution for the feather trade almost led to extinction, and the *Handlist* (1964) recorded that at the turn of the century "it was known to nest only in the Ellesmere area". Numbers have grown steadily since, and detailed surveys in 1955 and 1962 found 20 sites holding 33 pairs, and 28 sites holding 69 pairs, respectively. In 1965, 39 sites were found to hold 95 pairs (*SBR*). There are now 71 occupied tetrads and, as some of the larger waters hold more than one pair, a population of 150 to 200 pairs is suggested.

The presence of pairs on rivers indicates that the more suitable still-water sites are fully occupied, and further expansion may be limited to the river systems unless suitable additional waters are created. Nationally the population is still increasing, attributed to legislative protection from 1870 onwards and the post-war creation of additional habitat at gravel pits and reservoirs (*Pop. Trends* 1990).

These delightful birds may be threatened by the increased recreational use of water but they deserve our protection. Provided quiet corners are left they can breed successfully alongside fishing and boating activity although the wash from water-ski-ing can cause flooding of nests. High levels of disturbance on public waters can be a limiting factor although the breeding season extends into later summer, allowing them to try again if eggs are taken or the nest is destroyed.   *CEW*

BLACK-NECKED GREBE

Podiceps nigricollis

Status: **Occasional**

Tetrads with evidence of breeding

Confirmed	1 (0%)
Probable	0 (0%)
Possible	0 (0%)
Total	**1 (0%)**

JS

Recorded as an occasional passage migrant and winter visitor in seven of the years 1982–90 (*SBR*s), this spectacular and diminutive grebe is also one of Shropshire's rarest breeding species. No map is published and no details of locality or type of site are given here.

Forrest (1899), who referred to it as Eared Grebe, only mentioned two killed just beyond the border on Hanmer Mere, Clwyd, in 1864. The *Handlist* (1964) classed it as an irregular passage migrant, with only seven records between 1956 and 1963.

A pair has now been recorded at one site as follows:

Year	No. present	Total eggs	Total young
1985	1 pair	Not known	2
1986	1 pair	4	Not known
1988	1 pair	Not known	Not known

These are the first breeding records for Shropshire. The site was visited each year by a single observer, but no records were obtained in any of the other three years. However, the Black-necked Grebe is elusive and may possibly be overlooked, both at this site and elsewhere.

Breeding was first proved in Wales in 1904, England 1918 and Scotland 1930, with sporadic breeding in many counties since then. Throughout Britain and Ireland they were confirmed breeding in only five 10-km squares, probably bred in another two and possibly in four others (BTO *Atlas*, 1976). Recent records have come from four or five main colonies with occasional pairs elsewhere, the national population being 25–30 pairs (*Red Data* 1990).

Black-necked Grebes show an apparent preference for lowland eutrophic waters, but may desert well-established sites for no apparent reason. The Rare Breeding Birds Panel has recorded them breeding in central, eastern and southern England in recent years, with increasing numbers in suitable habitat, though some breeding attempts have been abandoned after human disturbance. The long-term trend is clearly upward (Spencer *et al.* 1990). Quite small waters may be used for breeding, and there is ample scope within Shropshire for expansion.

JS

GREY HERON
Ardea cinerea

Status: **Resident**

Tetrads with evidence of breeding

Confirmed	22 (3%)
Probable	23 (3%)
Possible	206 (24%)
Total	**251 (29%)**

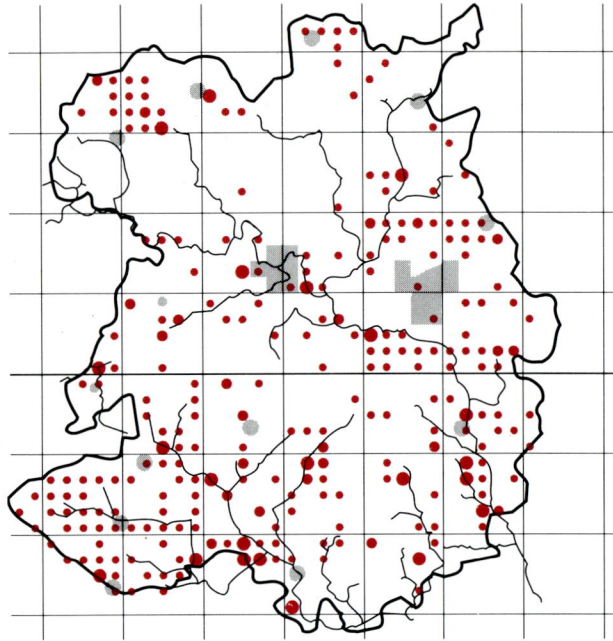

The Grey Heron is often encountered at the edge of rivers, streams and lakes, even resorting to upland pools and garden ponds to feed. It usually nests in small colonies in the tops of mature trees close to the feeding areas, although occasional nests have been recorded lower down in bushes. All the confirmed breeding records come from river valleys or are close to lakes. It is likely that a few small sites have been missed considering the scatter of records in the Clun valley and around The Weald Moors. The presence of colonies just outside the county at Aqualate, east of Newport, and Combermere, north-east of Whitchurch, may account for the scatter of records in these areas. Adults travel some distance to find food for the young and many of the smallest dots will refer to them or to non-breeders.

Most of the established heronries are well known and some have been occupied for many years. The largest, at Halston, dates back to at least the 1820s when up to 100 nests were noted. By the 1960s this colony had declined to about 40 pairs, probably due to the drainage of the marshes around the River Perry and associated loss of feeding grounds, and after the hard winter of 1962–63 only 14 nests were counted (*Handlist* 1964). Numbers continued to drop, possibly due to pesticide poisoning (Edwards *SBR* 1971), but after falling to less than 10 pairs in the early 1970s they began to increase again (Wright *SBR* 1985) until the hard winter of 1985–86 reduced the colony to 7 pairs (*SBR* 1986). Since then the population has grown both locally and nationally, due to legislative protection, a ban on pesticides and mild winters (*Pop. Trends* 1990) and by 1990 there were 28 pairs at Halston. The best-known heronry, which is at Ellesmere, increased from 8 pairs in 1985 to 25 in 1990. The nests are built on a small island clearly visible from the nearby road and in 1990 the RSPB set up a public observation hide attracting some 18,000 visitors.

Most of the other known sites are quite small, with three to twelve pairs, so a population of 100–120 pairs at twenty-two sites is likely. This compares with 67 nests at seven sites found in 1985 as part of a national survey (Wright *SBR* 1985). Some of the additional sites are of recent origin, presumably replacing those abandoned or destroyed, but a few may have been overlooked in 1985. Tree felling has caused the loss of several sites in recent years, and the new ring-road at Shrewsbury has narrowly missed taking out the only established heronry near the town, though the disturbance may still cause its desertion.   *CEW*

MUTE SWAN
Cygnus olor

Status: **Resident**

Tetrads with evidence of breeding

Confirmed	88 (10%)
Probable	41 (5%)
Possible	28 (3%)
Total	**157 (18%)**

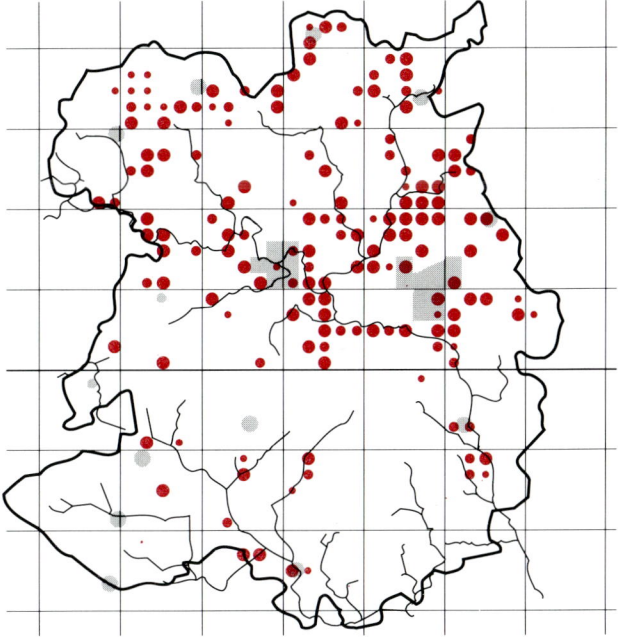

The Mute Swan is relatively easy to find once suitable habitat has been located, as both the birds and the nest are visible from some distance. Most pairs breed in the Severn valley and the north, wherever slow-flowing rivers, canals, lakes and pools are found. Even a small pool in a housing estate may be used if there is an island to offer security. Most of the nests along the Severn are on islands as the steep river-bank and the sudden changes in water level make the water's edge unattractive. In the south and west there is little suitable habitat and the few pairs present generally rely on man-made pools, with the occasional pair on the River Teme. Quite small pools may be used and if these are not marked on the map a site can be overlooked.

Breeding is quickly established by the presence of the nest or young cygnets which stay with the parents for several months. The presence of a pair with a well-grown brood is not always reliable evidence that they bred in that tetrad, as family parties may move some distance in search of better feeding grounds. A pair that nest regularly on a small pool in Newport walk the brood through the town centre, often with a police escort, to the canal.

Breeding was confirmed in eighty-eight tetrads, although some sites were not used every year. Some of the twenty-eight probable breeding records may refer to non-breeders or pairs trying to establish a breeding site. The final year of the Atlas coincided with a national census when 74 pairs were located in Shropshire, of which 52 pairs nested, and a further 152 non-breeders were counted. The number of nesting pairs can be compared with previous surveys — 78 in 1955–56, 72 in 1961, 29 in 1978, and 21 in 1983 (Campbell 1960, Eltringham 1963, unpublished BTO data for Shropshire) — showing a recovery now from the decline attributed to the ingestion of lead weights from fishing lines. The sale of these weights was made illegal in 1987 and their continued use is banned on many waters. The population increased during the Atlas period and by 1990 had risen to over 300, including 52–74 breeding pairs.

In urban areas they become very tame, but sometimes suffer from vandalism. Collisions with overhead wires will continue to cause many deaths and a mink has been seen to take young cygnets from a nest on the Teme despite vigorous defence by the adults (G.B. Thomas, pers. comm.). Improved survival of the young has led to the re-establishment of non-breeding flocks along the Severn valley. As they mature, the number of breeding pairs is expected to increase and the availability of suitable habitat may eventually become the limiting factor. *CEW*

GREYLAG GOOSE

Anser anser

Status: **Resident**

Tetrads with evidence of breeding

Confirmed	11 (1%)
Probable	14 (2%)
Possible	16 (2%)
Total	**41 (5%)**

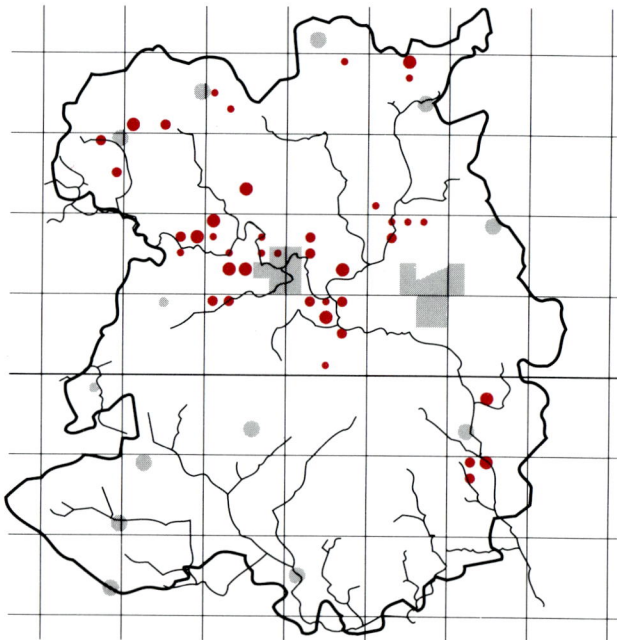

Wild Greylag Geese sometimes occur as winter visitors, but there is now a resident feral breeding population.

The wild population of Britain's only indigenous goose is restricted to northern Scotland. Wildfowlers deliberately created new colonies, initially in south-west Scotland in the 1930s, and then in 13 English and Welsh counties between 1961 and 1970 (BTO *Atlas* 1976).

The *Handlist* (1964) described it as a "non-breeding visitor in small numbers" with little doubt that some were escapes from collections. *SBR* records increased slowly, and the first confirmed breeding, a pair and six young at Bickley in 1969 "were no doubt feral birds, which are becoming more common throughout England largely owing to spread from the Wildfowlers Association Reserves, where the birds are bred for restocking" (*SBR* 1969). One such reserve (of the renamed British Association of Shooting and Conservation) at Nib Heath Pool is the local stronghold, and 42 (including young of the year) were counted in 1981, rising to 116 in 1989. Other sites with confirmed breeding are Shavington, Park Hall Camp, Marton Pool (Baschurch), Felton Butler, Alkmond Park Pool, Onslow, Venus Pool, Cranmere Bog and Dudmaston.

These favoured habitats are lowland meres and ponds, preferably with islands for nesting, and adjoining pasture, cereal or potato fields for grazing. The nest, usually on the ground near water, is constructed by the female in a sheltered hollow at the base of a tree, under a bush or in reed beds. The same site is frequently used in subsequent years. Eggs are normally laid in mid-April, but a replacement clutch may be laid if early egg-loss occurs. One brood is reared and breeding is usually confirmed by observing goslings still accompanied by parents.

The Greylag Goose occasionally hybridises with the Canada Goose and there is some interspecific competition between them for nest sites, as habitat requirements are similar.

Though numbers appear to have grown to around 150–300 individuals, only Nib Heath regularly has records of more than one to two broods, so a substantial part of the population must be non-breeding. An average of 2–3 pairs per occupied tetrad would give a population of around 50–75 pairs though, provided it can withstand competition from the larger Canada Goose, the Greylag Goose should continue to increase.   *MW*

CANADA GOOSE
Branta canadensis

Status: **Resident**

Tetrads with evidence of breeding

Confirmed	195 (22%)
Probable	78 (9%)
Possible	33 (4%)
Total	**306 (35%)**

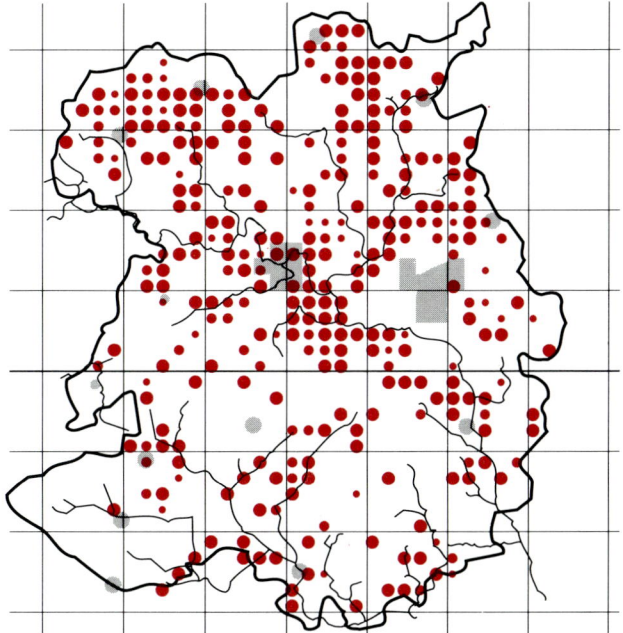

During the reign of Charles II the Canada Goose was introduced to Britain and a feral population became established during the 18th century. The date of arrival in

Shropshire is unknown but it was present on ornamental waters and ponds for many years prior to the 1860s (*Handlist* 1964).

The preferred habitat is lowland meres and pools not covered by vegetation, with islands for nest sites and gently sloping edges giving access to pasture for grazing. Fast-flowing rivers are usually avoided. It nests singly or colonially. The female builds the nest on the ground, usually within 30 metres of water, and often under a tree or bush or in rank vegetation for protection. Occasionally the nest will be in the open or some distance from water. Eggs are laid between late March and May but re-laying may replace early losses. One brood is reared, and breeding is usually confirmed by observing nests on islands, which may be conspicuous, or progeny being tended by parents.

Gregarious habits, the ability to adapt to human influence, and the extent of suitable habitat have enabled the Canada Goose to become one of the most common resident breeding waterfowl. Densities are higher in the north due to the greater number of suitable waters and it tends to be absent from some upland pools owing to the lack of cover.

The *Handlist* (1964) reported it absent as a breeding species from a large part of the south and south-east. The BTO *Atlas* (1976) did not record it at all in four 10-km squares in the same area, and only possible breeding was recorded then in two other 10-km squares in the north-east. There are now at least two confirmed breeding records in each of these six 10-km squares, so the range has increased considerably. Nationally the current rate of population increase is 8% per year (*Pop. Trends* 1990).

The sedentary and highly territorial behaviour of the Canada Goose creates regional sub-populations and limits the numbers of pairs that can occupy nesting sites. This natural control therefore arrests an even greater increase in the north Shropshire and south Cheshire sub-population. Experience of Atlas fieldwork suggests an average of two to three and a half pairs per occupied tetrad, giving a population estimate of between 550 and 1000 breeding pairs, plus a considerable non-breeding population.   *MW*

JS

BARNACLE GOOSE
Branta leucopsis

Status: **Occasional**

Tetrads with evidence of breeding

Confirmed	1 (	0%)
Probable	2 (	0%)
Possible	2 (	0%)
Total	**5 (**	**1%)**

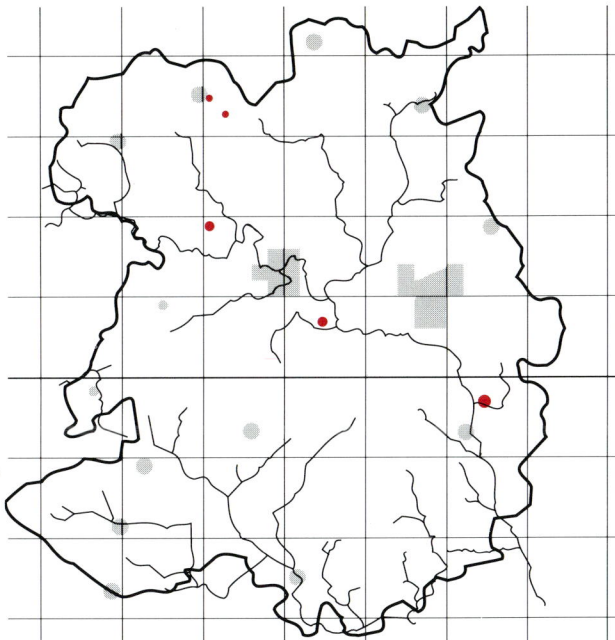

JS

The world population of Barnacle Goose, though growing, is only around 93,000. Extremely loyal to a few sites, only four breeding and three wintering areas are used. Of these, the Greenland population winters in western Scotland and Ireland, while those from Spitzbergen winter in the Solway Firth, and number around 30,000 and 12,000 respectively. Conservation and protection of these vital wintering areas puts an international obligation on Britain and Ireland (*Red Data* 1990).

Captive Barnacle Geese have bred in wildfowl collections for many years and escapes from them are almost certainly responsible for the small feral population now building up in Britain.

The *Handlist* (1964) described it as a "non-breeding visitor, rare" and there were only seven records in the 15 years after 1956, but from 1972 they were recorded annually (except in 1975). Very few were breeding-season reports until 1981 when one individual stayed at Colemere for most of the year. This mere held 4–6 for most of 1982–86, with a maximum of 8 in 1987, and provided the first confirmed breeding when a pair raised two young in 1984. In 1985, 6 were seen at Cranmere Bog, Worfield, and a pair raised two young there. Neither parent was ringed, suggesting that they too were not bred in captivity.

Subsequently these small flocks appear to have dispersed so the number of sightings, usually of ones or twos, has increased substantially and came from a dozen different waters throughout 1987. The probable breeding records relate to pairs at Venus Pool in 1987 and 1990, and at Nib Heath in 1989. The possibles relate to the Colemere flock, some of which were also seen at Ellesmere (*SBR*s).

The wild population migrates to breed on rocky Arctic coasts or cliff ledges, but the feral population appears to be sedentary, behaviour similar to that of the much larger and longer-established Canada Goose population. Barnacle Geese here are almost always associated with Canadas and appear to have the same habitat requirements, and some of the hybrid geese seen in Canada flocks appear to be Canada X Barnacle. It remains to be seen whether the currently small population, with the occasional breeding pair, can become established in the face of competition from its larger aggressive relative.

LS

SHELDUCK

Tadorna tadorna

Status: **Resident**

Tetrads with evidence of breeding

Confirmed	3 (0%)
Probable	9 (1%)
Possible	8 (1%)
Total	**20 (2%)**

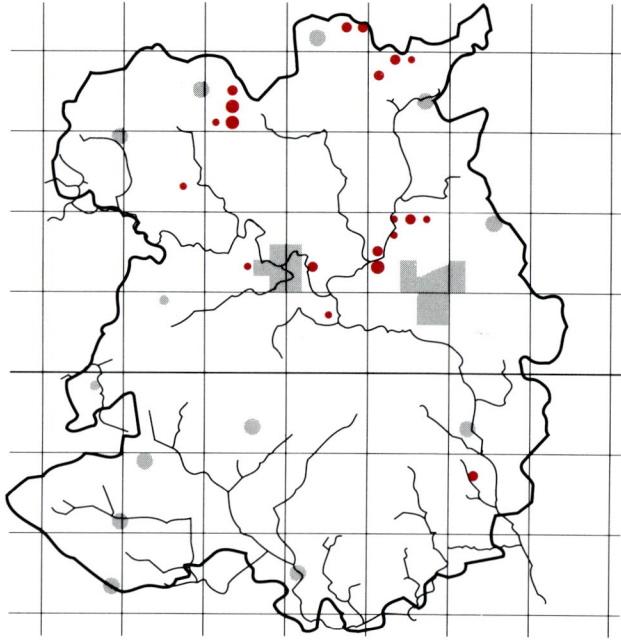

Shelduck first bred in 1963, one of seven duck species to start nesting here over the last 150 years, four of them since 1900.

Shelduck mainly inhabit the coast, but a considerable population increase during this century has been accompanied by a tendency to nest in small numbers in inland areas, including several land-locked counties since 1950 (*Pop. Trends* 1990).

The principal habitat requirement is shallow water in which adults and young can wade and dabble in search of invertebrates, especially small molluscs, as they do in the more usual estuarine environment. Rabbit burrows are the typical nesting place, but a wide range of sites may be used, including hay and straw stacks and tree holes.

The *Handlist* (1964) described the Shelduck as "mainly a passage migrant and non-breeding visitor" and this remains true today. Indeed, following the first confirmed breeding record at Allscott Sugar Factory in 1963, it was not until 1983, when a pair raised six young near Colemere, that breeding occurred again. Subsequently it has been confirmed in 1984 (twice, near Newton Mere, and at Wood Lane), 1985 (near Crose Mere), 1986 (near Wood Lane), and 1988 (at Allscott Sugar Factory). Probable breeding records have been reinforced by observation of juveniles in August at Venus Pool (1989, 1990) and Chelmarsh (1990) (*SBRs*).

Although breeding was confirmed in three tetrads, the records were from different years. Successful breeding is unlikely to have gone unrecorded and although the nine probable records may include some failed attempts, others will concern non-breeders, as Shelduck do not breed before their third calendar-year (*Red Data* 1990). Shelduck remain a rare breeding species and a maximum of three pairs is likely at present, with perhaps just one pair the norm. *TW*

MANDARIN
Aix galericulata

Status: **Regular**

Tetrads with evidence of breeding

Confirmed	3 (0%)
Probable	1 (0%)
Possible	0 (0%)
Total	**4 (0%)**

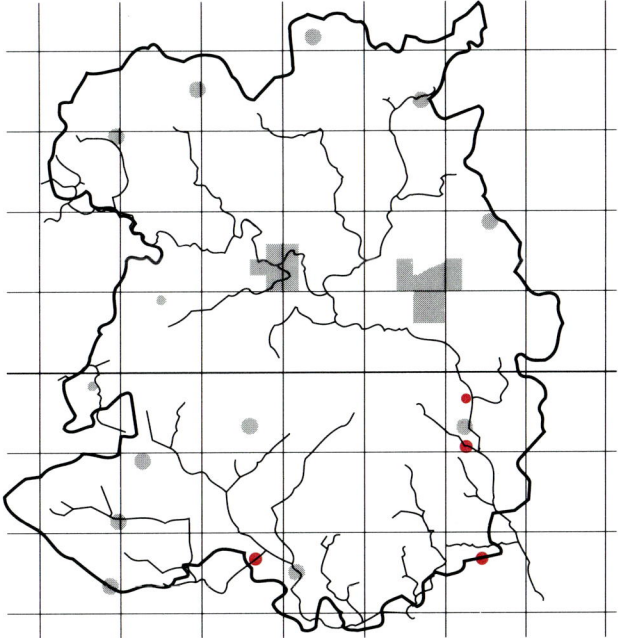

The beautiful plumage of the male Mandarin makes it popular with aviculturalists. A native of the Far East, it has established itself in the wild in England through escapes from collections. In Shropshire occasional records of individuals, usually males, were noted from 1977 onwards, mainly in the winter.

During Atlas fieldwork a small flock was located on the River Severn in the Quatford area, and breeding was confirmed when a female was seen on the river with four young. These birds are believed to be escapes from the former West Midland Bird Gardens which were less than a mile from the river at this point (*SBR* 1988). The flock currently contains at least a dozen birds. Of the other three records, two are close to the Severn and may have originated from the Quatford flock, whilst the fourth relates to a pair on the River Teme near Ludlow.

The Mandarin nests in holes in trees well above ground level, is secretive near the nest, and prefers secluded waters surrounded by deciduous woodland, so it may well be overlooked. The British population is still increasing (*Pop. Trends* 1990), and with a considerable spread of winter records along the Severn valley and around the meres in the north, it is likely that the current population of 3–6 breeding pairs will increase. *CEW*

GADWALL
Anas strepera

Status: **Regular**

Tetrads with evidence of breeding

Confirmed	1 (0%)
Probable	4 (0%)
Possible	2 (0%)
Total	**7 (1%)**

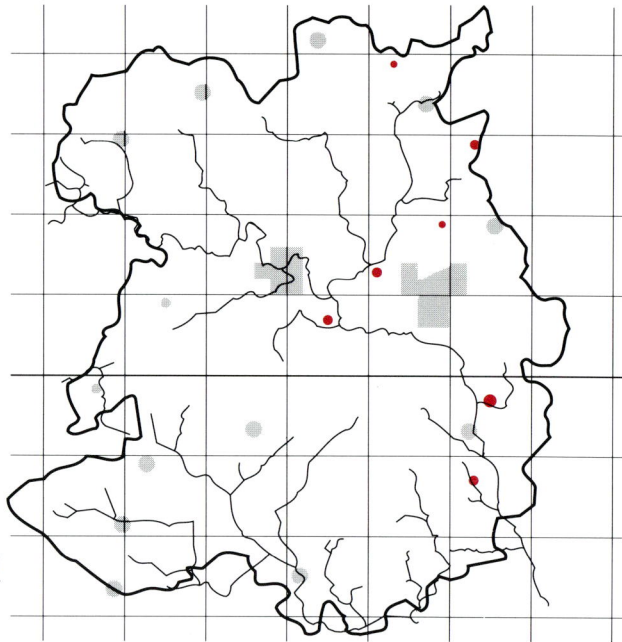

Gadwall did not breed in Britain until 1850, in Norfolk, when a pair were trapped during the winter, wing-clipped and turned down. East Anglia remains the main breeding area, although many further releases have led to the establishment of small populations in several other places, and it has been estimated that some 600 pairs now breed in Britain and Ireland (Fox 1988).

Nevertheless Gadwall have remained rare in Shropshire at any season, though they are most frequently seen in winter, and it was not until 1980 that breeding was finally confirmed, at Venus Pool. Subsequently breeding occurred at Shrewsbury Sewage Farm (1981), Priorslee (1984) and, during the Atlas years, at Chetwynd (1986) and Cranmere Bog (1988). Although the lake at Chetwynd is wholly in Shropshire, more than half the area of the tetrad is in Staffordshire, so this record does not appear on the Atlas map. Probable breeding records are from Allscott Sugar Factory and Venus Pool (1987), Knighton Reservoir (1988) and Chelmarsh (1990) (*SBR*s).

Breeding pairs choose shallow lowland waters surrounded by luxuriant vegetation in which the nest is well hidden. Gadwall are inconspicuous and, even when seen, females may readily be dismissed as Mallard. Nevertheless it seems unlikely that more than one to two pairs currently breed.

The national population continues to increase, despite the phasing out of releases, and Gadwall now breed regularly in Cheshire (*Pop. Trends* 1990). The Shropshire population therefore seems likely to rise.

TW

TEAL

Anas crecca

Status: **Resident**

Tetrads with evidence of breeding

Confirmed	12 (1%)
Probable	28 (3%)
Possible	21 (2%)
Total	**61 (7%)**

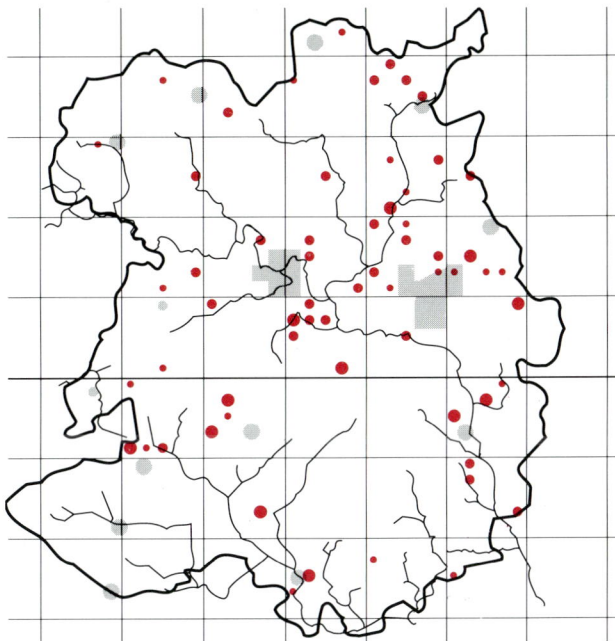

Although Teal are common winter visitors, they rarely nest. Lakes, rivers and marshes may be used for breeding, provided they offer both shallow water and plenty of emergent or other peripheral vegetation; but use of this habitat is only sporadic and the boggy pools of The Long Mynd are preferred. In five of the years 1981–90, including three of the Atlas years, one or other of these moorland pools produced confirmed breeding records. No other site was recorded as occupied in more than one season, apart from Venus Pool.

The meres (defined by Reynolds (1979) as extending from Ellesmere down to Berrington Pool, south of Shrewsbury) might be thought suitable, but are generally shunned; perhaps they offer inadequate sheltered habitat.

The only fully documented confirmed breeding records are for Betton Pool (Forrest 1934) and Alkmond Park Pool (*SBR* 1983), although the BTO *Atlas* (1976) also shows confirmed breeding in the 10-km squares SJ43 and SJ50, both of which include several of the meres.

The generally sporadic use of nest sites is well illustrated by comparing the present records with those for the BTO *Atlas* (1976). The latter recorded confirmed breeding in five 10-km squares, only two of which have provided repeat records for the present Atlas. Also at the well-watched Venus Pool, breeding has occurred only twice since 1968, in 1980 and 1983.

Proving breeding is not easy, as nests are hard to find, being sited in thick cover up to 150m from the nearest water, and the female and her brood usually remain well hidden. However, it would be optimistic to suppose that many of the probable records in fact concern breeding pairs, as passage birds and non-breeders will account for most of them. With this in mind, five or so pairs seems a sensible estimate of the present breeding population.

35

This contrasts markedly with the statement of Beckwith (1879): "the small wet bogs, so common throughout the county, are usually frequented in summer by a pair of ducks [Mallard], as well as a pair of Teal"; and that of Forrest (1908): "a good many pairs . . . nest on the more secluded pools and bogs". By contrast the summary in the *Handlist* (1964), "resident in small numbers", holds good today. *TW*

MALLARD
Anas platyrhynchos

Status: **Resident**

Tetrads with evidence of breeding

Confirmed	519 (60%)
Probable	147 (17%)
Possible	56 (6%)
Total	**722 (83%)**

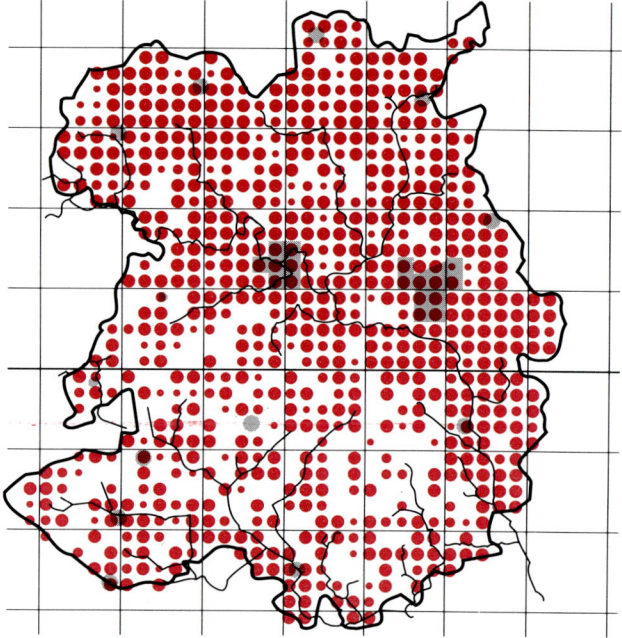

The widespread Mallard appears to be second only to Moorhen as the most abundant resident waterbird. Success is attributed to its ability to live alongside man, together with the large number of rivers, meres, pools and streams which provide countywide breeding habitats. It is absent only from some areas of upland which lack cover or slow-flowing stretches on streams, and other areas well away from suitable waters.

Courtship starts during early autumn and is observed discontinuously throughout the winter. The Mallard is catholic in choice of nest site and may be recorded up to 2 km from waters in rural areas. The duck builds the nest, usually in cover on the ground, or in the crown of pollarded willows; it readily accepts artificial nestboxes and baskets, and may become semi-domesticated, as at Whittington Castle adjacent to the A5 highway. The first young may be seen from late March but most hatch in May. One brood is reared but a replacement clutch may be laid after early loss, due perhaps to predation by corvids, stoats or weasels, or even after the demise of ducklings. Breeding is readily confirmed during the long season by sight of progeny which are usually accompanied by the female parent. Interbreeding with captive forms often occurs and hybrids are common at several places on the River Severn, and at Ellesmere.

Chicks are purchased, reared and released for shooting on many waters, notably at Fenemere and Marton Pool, Baschurch, as Mallard is the wildfowlers' favoured quarry. Winter feeding by the public, especially in Shrewsbury, Telford and Ellesmere, also helps sustain a greater population than would be supported naturally.

The average CBC density on farmland, equivalent to ten pairs/occupied tetrad (BTO *Atlas* 1976), may be reached in the north and east, with perhaps only one to two pairs in the south and west where the Mallard is much more scarce, giving an overall average of around 3–6 pairs, or 2000 to 4000 pairs in total. The recent ban on anglers' lead weights will reduce instances of poisoning and may result in a population increase in future.   *MW*

GARGANEY
Anas querquedula

Status: **Possible**

Tetrads with evidence of breeding

Confirmed	0 (0%)
Probable	0 (0%)
Possible	2 (0%)
Total	**2 (0%)**

Forrest (1899) referred to a nest found "near Shrewsbury, about 1888". This remains the only nesting record and Forrest otherwise regarded Garganey as "very rare". The *Handlist* (1964) stated: "passage migrant in small numbers".

The six Atlas years produced only three records, all of single males, at Chelmarsh (31 August 1986), Tong Lake (25 April 1987) and Venus Pool (29 May 1989) (*SBR*s).

Britain is on the edge of the Garganey's range and the national population is small, generally fluctuating around 40–60 pairs. Most are found in East Anglia and south-east England (*Pop. Trends* 1990), so it is not surprising that Shropshire is rarely visited, especially as the breeding habitat — shallow freshwaters and marshes — is uncommon.   *TW*

SHOVELER
Anas clypeata

Status: **Regular**

Tetrads with evidence of breeding

Confirmed	4 (0%)
Probable	11 (1%)
Possible	8 (1%)
Total	**23 (3%)**

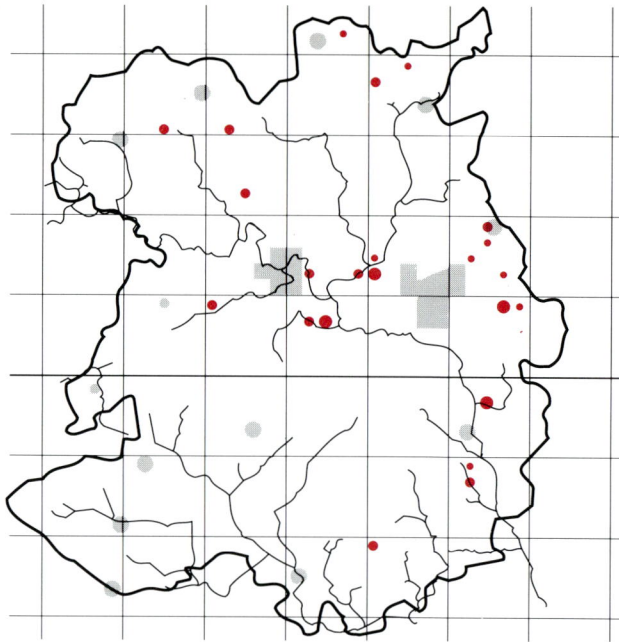

PW.

Parslow (1967) stated that "south-west of a line from the Cheshire Dee to Dungeness, probably the only places regularly supporting more than the occasional breeding pair are the Shropshire meres (where it has increased markedly in recent years) and the new Chew Valley Lake in Somerset". But prior to the late 1960s the Shoveler had been a rare breeding species in Shropshire and it quickly reverted to the same status.

The first breeding record, at "Hencote" (presumably Hencott, near Shrewsbury), was in 1869. It was followed by nesting records around the turn of the century at two further locations (Forrest 1908). Thereafter, breeding was not confirmed again until 1956, but subsequently numbers grew, until in 1968 probable or confirmed breeding was reported at nine sites (*SBR*). However, the BTO *Atlas* (1976) shows only one 10-km square with breeding confirmed and this reflects a rapid decline. Indeed over the fifteen years 1970–84 breeding was confirmed in only six years, with a maximum of only two sites reported in any one year.

Over the years 1985–90, breeding was confirmed in only four tetrads. Pairs are unlikely to be overlooked as the drake is very conspicuous, but breeding is difficult to prove unless ducklings are seen, and a few of the probables may have been breeding pairs. Nevertheless the Shoveler is clearly a rare breeding species, with none confirmed in some years, and more than three pairs in any one year exceptional.

Part of the reason for its scarcity is the lack of suitable breeding habitat — marshland and shallow eutrophic waters. The meres apparently proved suitable in the past but some change — possibly increased disturbance — may have rendered them unsuitable. However, eastern England is most favoured, and although Shropshire seems to have benefited from the major increase and spread that occurred in the first half of the century, it still lies on the edge of the range (BTO *Atlas* 1976). Fluctuations in status may therefore be expected.

TW

POCHARD

Aythya ferina

Status: **Resident**

Tetrads with evidence of breeding

Confirmed	5 (1%)
Probable	6 (1%)
Possible	16 (2%)
Total	**27 (3%)**

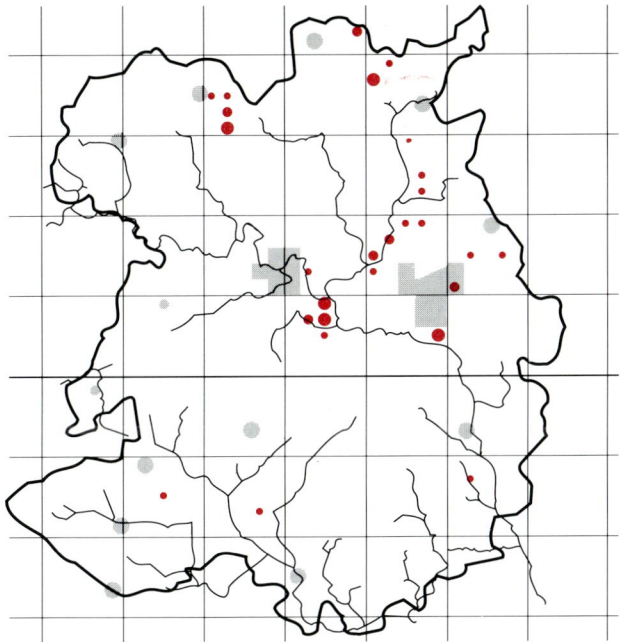

"Highly sporadic" is the only way to describe the incidence of breeding by Pochard up until 1980.

One nest was found at Tong Mere, near Shifnal, in 1875 and two at Whixall Moss in 1916. Thereafter, breeding was reported from "near Shrewsbury" in 1961 and 1963 and also, in the latter year, from "near Worfield", while the BTO *Atlas* (1976) shows confirmed breeding only from the Ellesmere group of meres. In 1980 young were seen at Venus Pool and subsequently breeding has occurred there in most years (Forrest 1908; *Handlist* 1964; *SBR*s). Nevertheless, the Pochard, although a common winter visitor, remains a very rare breeding species, as shown by the table below.

Pochard breeding records for Shropshire, 1985–1990

Year	Localities	Breeding confirmed	Breeding probable	Breeding possible	Maximum total
1985	7	2	2	4	8
1986	2	1	0	1	2
1987	6	2	0	5	7
1988	2	0	4	1	5
1989	8	1	2	5	8
1990	5	0	2	3	5

(Data supplied by the County Bird Recorder)

Pochard are relatively conspicuous and though nests are hard to find, broods are quite readily detected, so it is believed that this table gives an accurate assessment of the population, now estimated at 2–4 breeding pairs.

The most recent estimates of the UK population are in the range 180–395 pairs. Most sites are in eastern England, the stronghold being in East Anglia where breeding is long-standing. Expansion of the population westwards over the last 150 years has finally led, it seems, to more frequent breeding in counties like Shropshire, where formerly it was only highly sporadic (Spencer *et al.* 1990; Fox 1991).

Relatively shallow waters edged with dense vegetation are chosen for breeding, and there appears to be more of this habitat than is currently occupied, but the easterly bias in Pochard distribution may mean that the population will remain very small.   *TW*

TUFTED DUCK
Aythya fuligula

Status: **Resident**

Tetrads with evidence of breeding

Confirmed	83 (10%)
Probable	107 (12%)
Possible	27 (3%)
Total	**217 (25%)**

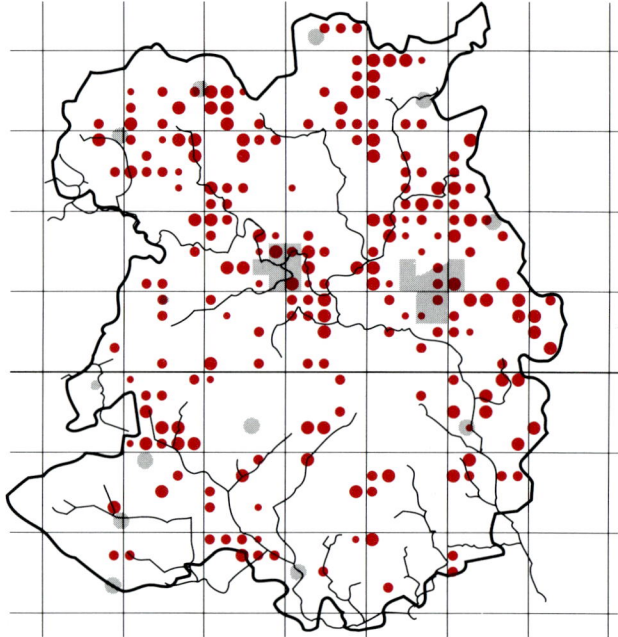

Tufted Duck was first known to breed in Britain in 1849, and the first breeding record here, at Hatton, Shifnal was not before about 1865 (Forrest 1908; BTO *Atlas* 1976). The *Handlist* (1964) estimated a breeding population of 35–40 pairs.

The normal nesting habitat is lowland lakes with some secluded bank-side cover. It is not unusual for sluggish flowing waters, such as the River Perry on Baggy Moor, to be used, but two habitat limitations noted in the BTO *Atlas* (1976) — normally no breeding above 400m, nor use of waters less than 1 ha in extent — no longer constrain Tufted Ducks in Shropshire. Boyne Water on Brown Clee at 455m is quite regularly used, while Norbury pool, near Bishop's Castle, is less than 0.2 ha in area, and is but one example of a number of small pools found holding broods during the Atlas years.

The meres and the artificial waters of the northern plain provide ample habitat of a more typical kind, reflected in a fair density of records. The thinner spread in the south is explained by the lower frequency of waters there.

Breeding is readily confirmed by the sighting of broods, but the map probably gives an exaggerated picture of the distribution of nesting Tufted Ducks in any one year, as some sites are only used sporadically, while summering non-breeders give rise to some probable breeding records.

However, some tetrads encompass more than one occupied water, and several pairs may nest at a single site, with up to three broods at some (*SBR* 1990) and five broods at Shrewsbury Sewage Farm (*SBR* 1989).

Working from the assumption that breeding takes place in some 80 tetrads in any one season and on average two to three pairs breed per tetrad, a Shropshire population of 160–240 may be inferred. This is approximately five times the figure for 1964, compared with the reported trebling of the national population between 1963 and 1983 to 7000–8000 breeding pairs (Owen *et al.* 1986). This growth is reflected in the occupation of high and small pools apparently unoccupied 20 years ago.   *TW*

GOOSANDER
Mergus merganser

Status: **Occasional**

Tetrads with evidence of breeding

Confirmed	2 (0%)
Probable	3 (0%)
Possible	2 (0%)
Total	**7 (1%)**

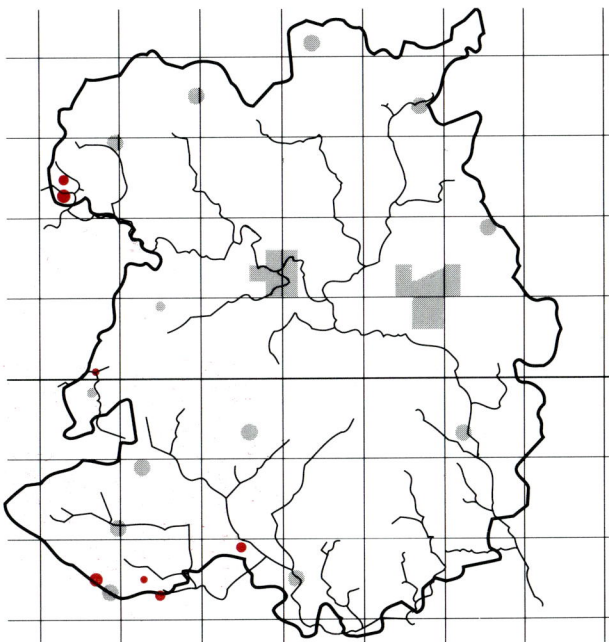

Long known as a winter visitor, Goosanders were not confirmed as breeding in Britain until 1871, in Perthshire and Argyll. Since then they have spread steadily, reaching northern England by 1941, Wales in 1970 and Devon in 1980 (*Pop. Trends* 1990).

The *Handlist* (1964) described Goosander as a "non-breeding visitor, regular from November to early April". Prior to 1948 numbers were small but since then they have built up, and a wider range of sites has been visited over the winter months. Breeding was first proved in 1987, when young were seen on the River Tanat, a tributary of the River Vyrnwy, and again in 1989 on the Teme. Other rivers which have generated breeding-season records are the Onny, Redlake and Camlad.

Goosanders are wary and the more conspicuous drakes leave British rivers in late May or early June (Little & Furness, 1985) so some may have been overlooked. Nevertheless the breeding population is unlikely to exceed five pairs.

Clear, fast-flowing rivers form the typical breeding habitat. Good numbers of fish are required, as are suitable nest holes in trees or rocks within a kilometre or so of the river. Few rivers meet these requirements, although the Ceiriog, Dee, Vyrnwy and Clun could provide sites in the future. In addition small tributaries may be used, broods being taken downstream to the main river (S. Carter, pers. comm.). Therefore even a river as small as the Redlake, narrow enough to be jumped where Goosander was recorded, is potential breeding habitat.

Whilst breeding Goosanders are likely to remain rare, birdwatchers in the south and west should nevertheless keep their eyes peeled, an exciting new breeding species may be just round that bend in the river!   TW

RUDDY DUCK
Oxyura jamaicensis

Status: **Resident**

Tetrads with evidence of breeding

Confirmed	13 (	1%)
Probable	27 (	3%)
Possible	13 (	1%)
Total	**53 (**	**6%)**

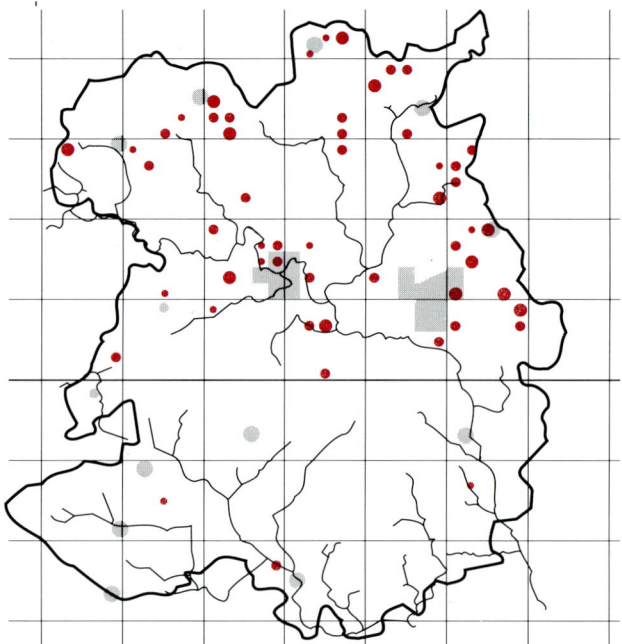

In 1936 the North American Ruddy Duck nested for the first time in Britain, at Walcot, Lydbury North, Shropshire — in a wildfowl collection (Anon. 1938).

In 1948, three pairs were brought from North America to Slimbridge, Gloucestershire. They bred in 1949 and subsequently a steady trickle of birds escaped. Feral breeding was first reported in 1960 and Shropshire was one of the earlier counties to be colonised, when a pair bred at Crose Mere in 1965 (Hudson 1976; *SBR* 1966).

Breeding habitat is still water, with some shallow water and aquatic vegetation, including emergent plants for nesting cover. North Shropshire offers a number of such waters but few records come from the south, where suitable conditions are much more scarce. A confirmed breeding site of particular note is Llyn Rhuddwyn on the Welsh border, west of Oswestry, at 335m.

Male Ruddy Ducks are colourful and obvious and, because of this, few occupied sites will have been overlooked. By contrast females are small and dowdy; they nest in thick cover late in the season, and broods often do not hatch before July, August or even early September. Consequently some breeding pairs will have been recorded as probables, but other probables may have been non-breeding birds, as they do not normally nest until two years old (Ogilvie 1975).

Most sites hold only one breeding pair but five nests were found at Crose Mere in 1981, and some tetrads have more than one occupied water. Assuming firstly, that breeding took place in half of the probable tetrads and secondly, that an average of 1.5 pairs bred in these and in confirmed tetrads, the population may be in the order of 40 pairs.

Pop. Trends (1990) suggests that the really dramatic increases of the last 30 years may be over, but some further expansion seems likely, and a breeding season sighting at Walcot, Lydbury North, in 1988 encourages the thought that, almost 60 years on, a wild pair could yet breed there.   *TW*

MARSH HARRIER
Circus aeruginosus

Status: **Rare**

Tetrads with evidence of breeding

Confirmed	1 (0%)
Probable	0 (0%)
Total	**1 (0%)**

Elsewhere in Britain Marsh Harriers breed almost entirely in extensive reed beds, though in recent years a few have nested in cereal fields. Previously widespread, they had disappeared from this country by the end of the 19th century following persecution and habitat drainage. They became re-established in the 1920s, but the slowly growing population was hit by pesticides in the 1960s. Expansion began again in the 1970s and by 1988 there were over 75 nests in Britain, mainly in East Anglia (*Red Data* 1990).

Recorded in the *Handlist* (1964) as a "non-breeding visitor, rare", there had only been four records since 1955 when fieldwork for the Atlas began: in May 1960, May and August 1980, and August 1984. A juvenile was recorded in September 1985 and then in 1988 a pair built a nest in a field of Italian ryegrass and laid at least two eggs. The grass was cut for silage but the farmer agreed to leave an area uncut around the nest. Apparently the eggs did not hatch and certainly no young were raised. The pair did not return the following year. An immature was recorded in September 1989, and an immature female in May 1990, both near Chelmarsh.

The confirmed breeding record was the first this century and has been fully documented (Wright *SBR* 1988), although the site remains confidential. Unless the British population expands considerably and nesting away from reed beds becomes common it is unlikely that the Marsh Harrier will become established in Shropshire.

CEW

HEN HARRIER
Circus cyaneus

Status: **Probable**

Tetrads with evidence of breeding

Confirmed	0 (0%)
Probable	1 (0%)
Total	**1 (0%)**

The silver-grey male and "ringtail" female are apparently so different they were once considered to be separate species. Though previously widespread, by the turn of the century they were restricted to the Orkney Islands and Outer Hebrides as a result of destruction of moorland and persecution by farmers and game-keepers. A slow expansion southwards started in the 1930s, reaching northern England and Wales by the 1960s, as they took advantage of young conifer plantations, a new habitat not controlled by gamekeepers. However, this opportunity is short-lived, as the trees rapidly grow too tall to allow nesting and hunting. The British population peaked in the 1970s and is now around 500 pairs. The recent decline has been caused by loss of habitat through continuing and maturing afforestation, and illegal persecution on moors managed for grouse shooting (*Red Data* 1990).

Prior to a small breeding population becoming established on the Welsh moorlands, Hen Harriers were rare in Shropshire, though "apparently two pairs nested in the south-west of the county in 1923" (*Handlist* 1964). Now a few records are received each year, and in 1988 a pair were seen holding territory just inside the border, but the tetrad is not included in the Atlas as it is predominantly in Wales, and breeding was not confirmed (*SBRs*).

Hen Harriers grace open moorland, especially where a strong growth of mature heather provides cover for the ground-nest and habitat for prey such as grouse, small birds, rabbits and other rodents. Several western areas may contain suitable breeding habitat, though it is likely to be marginal: grazing sheep may prevent the growth of sufficiently thick heather and trample any nests; the main nest-predators, foxes and crows, are common; and Shropshire is still a black spot for illegal persecution of birds of prey. However, if new legislation making landowners responsible for the actions of their gamekeepers provides effective protection for Hen Harriers in their prime habitat elsewhere, they are likely to spread from the grouse moors, and may perhaps become established here.   *LS*

GOSHAWK
Accipiter gentilis

Status: **Resident**

Goshawks were probably at one time widespread, but persecution caused extinction in England by the 1890s (*Red Data* 1990). Forrest's accounts (1899, 1908) made no mention of them but a Goshawk was one of nine rare birds of prey "obtained on the Longmynd between 1848 and 1857 by Rev. H. O. Wilson" (Forrest 1909). Thereafter Forrest (1930) noted that one escaped from Willey Hall (near Broseley), and the *Handlist* (1964) mentioned a pair nesting at Hawkstone Park (near Hodnet) in 1951: "They were robbed twice and the female was shot after laying three eggs in her third nest". She was recognised as an escaped bird. The *Handlist* (1964) could only report two other single sightings of individuals in 1959 and 1960. But a change was soon to occur.

A revival of falconry occurred in Britain in the 1950s. Goshawks had always been favoured and the sport's growth led to the establishment of a new feral population through the proliferation of escapes and some which were purposely released. Shropshire was one of the earlier counties involved, breeding being first proved in 1966. Subsequently it has occurred annually and Goshawks are now established residents in small numbers at several locations, although they are often overlooked, being elusive and secretive. Colonisation has been aided by coniferous afforestation because, although Goshawks will nest in small deciduous woods down to 3 ha in size and hunt in open country as well as woodland, productivity is highest in the larger, usually coniferous blocks, where disturbance is often lower (Petty 1989).

In spite of statutory protection, Goshawks are relentlessly persecuted by some keepers, falconers and egg collectors; birdwatchers can also cause illegal disturbance at or near the nest, or inadvertently alert persecutors. Nationally the population is at least 200 pairs, but "between 1979 and 1989, 39 adults were known to have been killed, and eggs or young were deliberately destroyed at seven nests" (RSPB 1991). Therefore no distribution map is shown nor is any indication given of the numbers of tetrads with evidence of breeding.

Several areas have suitable habitat, and favoured prey such as pigeons and corvids abound, so further colonisation can be expected where persecution and disturbance are absent.   *TW*

SPARROWHAWK

Accipiter nisus

Status: **Resident**

Tetrads with evidence of breeding

Confirmed	147 (17%)
Probable	129 (15%)
Possible	321 (37%)
Total	**597 (69%)**

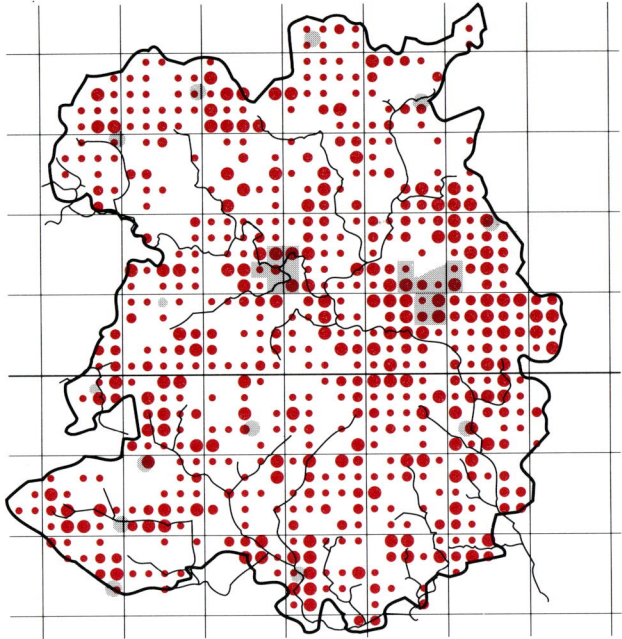

The dashing little male Sparrowhawk, grey-backed and rufous-breasted, is always guaranteed to provide an uplift to a day of "atlasing".

Its changing fortune over the past 30 years has been well documented. After an alarming decline in the mid-1950s, as the use of agrochemicals became prevalent, it steadily increased through the early 1960s as restrictions were placed on the use of aldrin, dieldrin and heptachlor. The *Handlist* (1964) described it as "now scarce, formerly quite common" and it was still not found in the main central agricultural areas during BTO *Atlas* fieldwork in 1968–72. Densities have been at or close to maximum since 1978, indicating recovery to be virtually complete (*Pop. Trends* 1990).

They now breed wherever there are woodlands, from blanket coniferous plantations in the south and south-west to quite small copses in agricultural country. They take readily to conifers and even in mixed woodland often prefer larch for the nest tree.

Territories are best located by noting old nests, as pairs remain faithful to an area, or by watching for soaring display over woods in spring. Plucking posts also provide clues to the vicinity of a nest. Confirmation of breeding is difficult, and is usually obtained through the infrequent observation of adults carrying food to the young: only the male when nestlings are small, and both parents later. Noisy nestlings may also betray a nest in the later stages. The number of occasions when a single Sparrowhawk is merely glimpsed is reflected in the high number of possible records.

Though a nesting pair may carry food through neighbouring tetrads, this may be offset by favoured tetrads containing more than one nest. Newton (1986) stated that in Britain densities of 14–96 pairs per 10-km square have been found, depending largely on the ratio of wooded to open terrain. Within these wide limits, one to three pairs in all tetrads recorded would give a population of 600–1800 breeding pairs.

There are suprising gaps on the map. A close correlation with a woodland species such as Great Spotted Woodpecker could be expected, but comparison shows Sparrowhawks to be recorded relatively sparsely, particularly in many southern areas. Almost certainly they are under-recorded owing to their elusive habits.   *JS*

BUZZARD
Buteo buteo

Status: **Resident**

Tetrads with evidence of breeding

Confirmed	174 (20%)
Probable	128 (15%)
Possible	123 (14%)
Total	**425 (49%)**

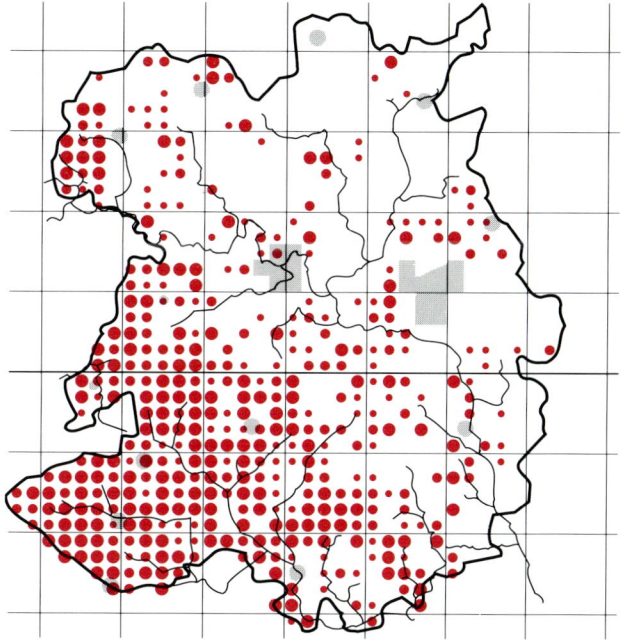

With deeply slotted wings and widespread tail, the largest common raptor is a familiar sight as it soars effortlessly on a rising thermal, making it a popular choice for the SOS emblem.

It breeds in greatest numbers in the woods of the south-west, with the range spreading across the south to the Clee hills and northwards along the western border to the Oswestry uplands. Thus it is present throughout the highlands (Map 3). Away from this habitat pairs breed in undulating country with scattered woods. The blank areas are mostly prime arable land which fails to supply satisfactory nesting woods and hunting needs.

The large size, soaring display and loud mewing call make territories easy to find in the spring, and nesting woods can be located even from a distant vantage point. Subsequent visits confirm breeding as adults carry food to the nest and later juveniles join parents in flight. Even so they can be remarkably elusive for such large birds, and some of the probable records should doubtless show confirmed breeding. The majority of possible records indicate non-breeding immatures and prospecting adults.

The Buzzard has been the subject of previous studies. The *Handlist* (1964) recorded an increase from a few pairs in the Clun Forest at the turn of the century to a peak in 1954, when a BTO survey located 74 pairs in a study area, which suggested a total population of 100 pairs. From that date the outbreak of myxomatosis almost eliminated the rabbit population, the Buzzard's staple diet, resulting in a decline to "about a quarter of their previous strength" (*Handlist* 1964). Recovery was slow due to the adverse effect of organochlorine residues as the use of chemicals increased. A more recent BTO/SOS survey (1983) concluded that around 80 pairs bred (Sankey *SBR* 1983). Further ground has been gained, especially in the east, but also due to infilling in more favoured areas. Breeding has now been confirmed in five 10-km squares where there were no records for the BTO *Atlas* (1976), and in a further nine where only probable or possible breeding was established then, reflecting a national increase over this period of 50% (*Pop. Trends* 1990). Though a few tetrads contain two pairs in some years, breeding does not occur every year in every occupied tetrad, suggesting an upper limit of 300 pairs.

Prospects for continuing expansion would be good were it not for irresponsible and illegal persecution. Unfortunately Shropshire is a black spot for the poisoning of raptors, and the Buzzard is the main victim (RSPB 1991). At the time of writing an investigation is in progress regarding a particularly distressing incident of a pair poisoned by alphachloralose on rabbit bait on an estate south-east of Shrewsbury in 1990. Several birds injured by shooting have been taken in for treatment during recent years (B. Williams RSPCA, pers. comm.), but those killed outright probably remain undiscovered. Another factor limiting expansion may be a poor rate of breeding success, due to the rabbit never regaining former numbers following the original outbreak of myxomatosis. The national annual breeding success averages less than one young fledged per breeding pair (*Pop. Trends* 1990), perhaps leaving insufficient young birds to initiate expansion of the range into new areas.   JS

KESTREL
Falco tinnunculus

Status: **Resident**

Tetrads with evidence of breeding

Confirmed	174 (20%)
Probable	174 (20%)
Possible	370 (43%)
Total	**718 (83%)**

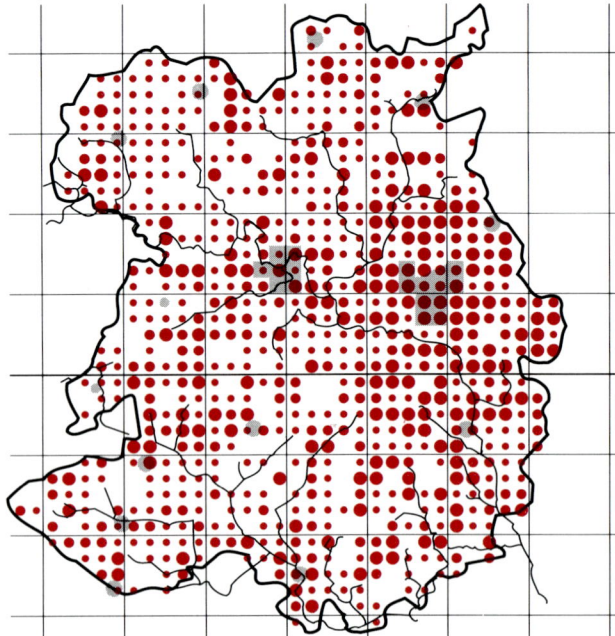

The Kestrel's habit of hovering makes it the easiest raptor to identify, and use of the wide grass verges of major roads allows easy observation.

Widespread and numerous, it may be outnumbered locally by Sparrowhawks in well-wooded areas, and is almost matched by them in the west. Also, it is surpassed by Buzzards in the south-west. By far the majority of confirmed and probable breeders are in the east, but the reasons are not obvious. Four factors which may favour the east are: field voles are the main prey and may be more numerous; young Kestrels are often fed on newly fledged birds of open country; there are more roadside verges; and urban areas, including parks and the grounds of schools, hospitals and factories, are readily used. Part of the imbalance may also be due to the higher number of observers in the east. Confirming breeding can be very difficult, although the ease with which Kestrels can be seen hunting has led to the high proportion of possible breeding records. Many of these undoubtedly bred.

They suffered losses from organochlorine pesticides, especially in the cereal farmland of eastern England, but this coincided with a stronger trend towards a national increase, including an expansion into suburban areas (*Pop. Trends* 1990).

Kestrels breed mainly in hollow trees, the nests of other species, on quarry ledges and in nestboxes, and an absence of suitable sites will limit the population density. The BTO *Atlas* (1976) gave an average of 75 pairs per 10-km square based on CBC figures, but there has been a drop of 30% in the index since then, and this "recent decline at national level is largely due to decreases in western England and in Wales" (*Pop. Trends* 1990). This regional variation is also reflected in the imbalance in the distribution here, and while there may be 2–3 pairs per occupied tetrad in eastern Shropshire, there are far fewer than this in the west. One to two pairs in each tetrad recorded would give a population of 700–1400 pairs.

Numbers may be subject to cyclical fluctuations, as with other raptors and owls that feed mainly on small rodents. Provided there are sufficient nest sites, there is room for expansion by infilling in years of high prey density.   *JS*

MERLIN
Falco columbarius

Status: **Resident**

Tetrads with evidence of breeding

Confirmed	6 (1%)
Probable	1 (0%)
Total	7 (1%)

It is likely that this tiny and elusive falcon, seen at its best in dashing flight just above the heather in open country, has never been a common sight in Shropshire.

During the Atlas period two areas were used for breeding, although they covered six tetrads. One site has been used for many years, and results from this traditional area were:

Year	Evidence of breeding
1985	One pair raised three young.
1986	Two nests raised three and one young. It is thought that one male bred with two females, the nests being only 600m apart. Two females were seen in the area on three occasions, but never more than one male. (Wylie, pers. comm.).
1987	One pair probably breeding, no juveniles seen.
1988	One pair raised four young.
1989	Confirmed breeding, perhaps two pairs, but no juveniles seen.
1990	Probably breeding, perhaps two pairs, but no juveniles seen.

In 1989, a single male and female were also seen in separate suitable breeding habitats.

Merlins have probably bred annually over many years, but not always successfully. Of two nests wardened by the RSPB in 1984, one was deserted by the female. It may not be entirely coincidental that this site was known to, and visited by, many birdwatchers. The second pair, more remote and undisturbed, raised four young. In 1985 a pair abandoned the first site chosen, almost certainly due to unnecessary close attention from birdwatchers, but fortunately resettled some distance away.

Locating a territory is difficult, as pairs indulge in aerial display infrequently. Males bring food to the female incubating or brooding young at lengthy intervals, making the late nestling or early fledgling period the easiest time to confirm breeding. Territories may also be betrayed by aggression towards intruders, especially the crows and other raptors. Old crows' nests are used, and on at least two occasions grey squirrels have been seen to climb the nest tree, to be driven away by vicious, persistent attacks from the vigilant female.

Nationally, Merlins have declined steadily from the turn of the century, with acceleration since 1950. Whilst other raptors recovered from losses due to agrochemicals, this species, probably picking up the poisons in lowland winter quarters, has never done so, and the British population is estimated at only 550–650 pairs (*Pop. Trends* 1990). The reasons are not fully understood. Forrest (1899) listed it as occurring "not unfrequently" (*sic*), adding that it rarely bred but a nest was found in 1896. The *Handlist* (1964) stated that at least two pairs bred annually in one area and had done so for many years, and they had been noted in another area but breeding had not been proved. The present Shropshire situation is an improvement since then, as the discovery of perhaps two pairs apparently flourishing in a hitherto unknown area, made up largely of atypical habitats, alters the county situation considerably.

There has been no evidence of expansion within the traditional area, even though several sites satisfy the prime breeding requirements of dry heather moors, rock outcrops and old crows' nests. Disturbance from a wide range of outdoor recreational activities, and possibly a high rate of infant mortality, may be largely to blame. The population is not likely to exceed six breeding pairs.   *JS*

HOBBY
Falco subbuteo

Status: **Summer visitor**

Tetrads with evidence of breeding

Confirmed	4 (0%)
Probable	8 (1%)
Total	**12 (1%)**

Resembling a miniature Peregrine with grey upper plumage and bold black moustachial stripe, the sight of this dashing falcon in pursuit of hirundines was uncommon until recent years. The first returning migrants are rarely noted before the final week of April.

Long regarded as typical of lowland heath and downland, it is now known that more Hobbies breed in Britain than previously suspected, using farmland and woodland habitat in addition to the traditional areas (*Pop. Trends* 1990). Favoured sites are agricultural areas with scattered trees where old Carrion Crows' nests are used.

During the Atlas period three pairs bred successfully in widely separated tetrads, and pairs were seen in four other suitable breeding areas. Singles were seen in over 50 tetrads, mostly in a large central area, and in the north, north-east and south-east. Although they were probably hunting, breeding in these areas is a possibility, and the *Handlist* (1964) referred to a breeding site close to the Herefordshire border in 1950–54, and two other "old nesting haunts". Hobbies may have been present for some years but overlooked, although a significant uplift in sightings from 1984 suggests a genuine rise in population, as they appear to have increased and extended their range from about 1976 (*Pop. Trends* 1990).

The first SOS confirmed breeding record was in 1983, but it is still uncommon and vulnerable to egg collectors, so locations of nest sites are not published. The population could be up to eight breeding pairs.

Proving breeding is difficult, as parents hunt at great distances from the nest. Display may betray territory shortly after arrival, but confirmation is most likely to come from noisy adults reacting to an intruder near the nest when the young are at late nestling or early fledgling stage. Two nests were discovered by Atlas workers, and the other was reported by a gamekeeper. Information from non-birdwatchers can be of great importance, but misidentification can be a problem. An Atlas worker discussing one site with a farmer was told they had nested the previous season, but were thought to be Kestrels. A pair was present at one site throughout the Atlas period, using four nest trees, all within a 100m radius, but straddling two tetrads. Local information indicated this to be a new site in 1985. Yearly results from this one site were:

Year	Evidence of breeding
1985	Confirmed breeding, outcome unknown.
1986	Two young fledged.
1987	Pair present, the female thought to be a first summer bird. They will mate with older males, often unsuccessfully (*BWP*). No nest established.
1988	Pair present, and nest defended, but breeding unsuccessful.
1989	Single young fledged.
1990	Three young fledged.

JS

PEREGRINE
Falco peregrinus

Status: **Regular**

Tetrads with evidence of breeding

Confirmed	2 (0%)
Probable	0 (0%)
Total	**2 (0%)**

This popular falcon, with its commanding presence and renowned "stoop", is rapidly becoming a familiar sight to Shropshire birdwatchers.

During the fieldwork period two pairs bred. One nest, discovered in 1987, was the first officially reported to the SOS. Results were:

Year	Site	Breeding evidence
1987	Site A	Two young hatched, one fledged.
1988	Site A	Three eggs laid, two hatched, one fledged.
	Site B	Three eggs hatched, nestlings ringed, at least two fledged.
1989	Site A	Present at site, but breeding not confirmed.
	Site B	Three eggs laid, one young hatched, ringed and fledged.
1990	Site A	At least one young fledged.
	Site B	Present at site, but breeding not confirmed.

Throughout the Atlas period single Peregrines were also seen in other suitable locations. No map is published as they still face illegal persecution by killing, or theft of eggs and chicks. In 1991 four eggs were laid at both the above sites, but one clutch was stolen. Nationally an average of 45 eyries per year were robbed from 1981–88 (*Red Data* 1990). Unfortunately the adults tend to be very noisy around the breeding site, attracting attention to a location that may otherwise remain undetected.

In the pre-war decade 850 territories were known in Britain, but numbers had plummeted to an all-time low of 360 pairs by 1962, many of which failed to breed. There may be 1000 pairs today. The reasons for this wide fluctuation: official culling during the Second World War; subsequent recovery; the crash due to organochlorine pesticides; and, most recently, the build-up to the present healthy position, are well documented (*Pop. Trends* 1990). In Shropshire few were seen until the early 1980s, but from 1983 there came a sudden and sustained increase in records, including many breeding season sightings (*SBRs*). There are now two pairs plus a number of non-breeding adults.

Past breeding history is sparse. Forrest (1899) wrote "No recent instance is known of its nesting here" and the *Handlist* (1964) recorded it as a "Non-breeding visitor, rare". Ratcliffe (1980) referred to "a reputed former breeding place . . . in the uplands of western Shropshire". Hardy (1970, 1971) published news of three nests in 1970, at least one of which hatched young again the following year. Another discontinued breeding due to construction work in the area. These were all different sites from those used now.

If left free from persecution and disturbance, expansion would be likely. Several suitable sites could attract maturing young from the recent successful nests, and others from over the Welsh border.

JS

RED GROUSE
Lagopus lagopus

Status: **Resident**

Tetrads with evidence of breeding

Confirmed	8	(1%)
Probable	5	(1%)
Possible	3	(0%)
Total	**16**	**(2%)**

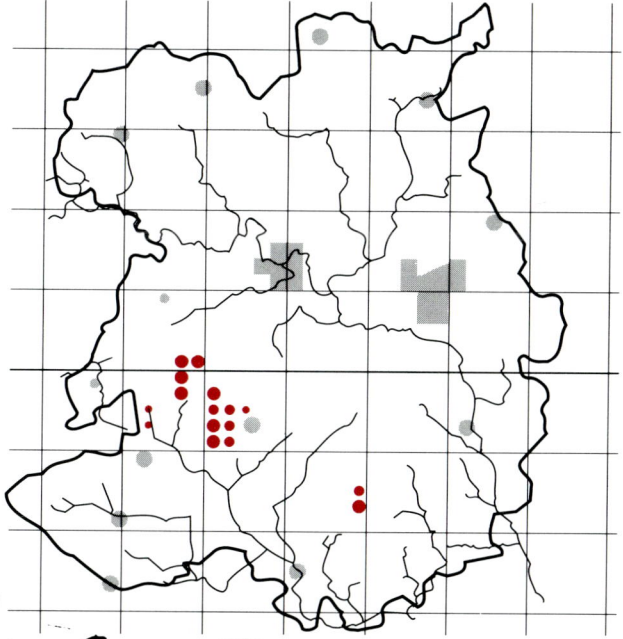

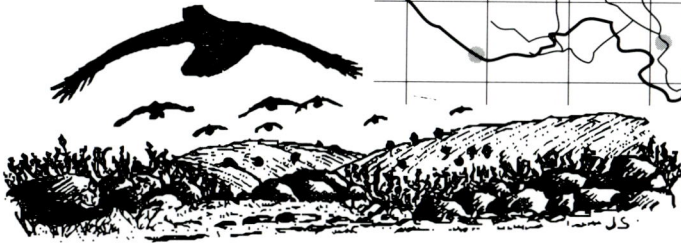

The habitat of Red Grouse is moorland dominated by common heather, its principal food. The strict habitat requirements, distinctive "go-back, go-back" call and characteristically swift and tilting flight mean that Red Grouse are easy to locate and identify. Nests are not easily found but chicks and recently fledged young are more readily encountered, making breeding relatively easy to prove at all the currently known haunts — The Long Mynd, the Stiperstones and Brown Clee.

Eyton (1838) referred to Red Grouse as "common on the Stiperstones" but Forrest (1899) stated that "careful enquiries" showed "few, if any" in Shropshire before 1840, when two pairs from Yorkshire were released on The Long Mynd. Forrest was well aware of Eyton's publication and the reason for the inconsistency between these two authorities is unknown. By the turn of the century Red Grouse were, according to Forrest, "plentiful" on The Long Mynd, Clun Forest and Clee hills. The *Handlist* (1964) gave The Long Mynd and the Stiperstones as the main localities, adding Stow Hill (near Knighton) as a surviving location and Brown Clee and Black Mountain (near Anchor) as former ones. Beckwith (1879) recorded 47 brace shot in a day on The Long Mynd and Forrest (1899) up to 90 brace. As recently as 1975 the

53

season's bag there was as high as 113 brace, but 32 brace shot in 1990 is the highest tally over the last decade. Records for the Stiperstones show 115 brace shot in 1911 but since the 1960s the maximum has been only 15.5 brace in 1988 (data from local game books).

These declines reflect national trends. Habitat degradation and loss, and relaxation of predator control, are regarded as key factors elsewhere and will have been influential here. Disease and cyclical fluctuations are both well known as factors affecting numbers, but they are most unlikely to have been the cause of the gross and persistent decline that has occurred and for which there is no adequate explanation.

Current population estimates for The Long Mynd and the Stiperstones are 50–75 and 15 breeding pairs respectively (National Trust and English Nature staff) and five pairs were reported from Brown Clee in 1989 (*SBR*). Birds are occasionally seen on Heath Mynd (south-west of Stiperstones) but breeding has not been proved recently. Two were seen on Titterstone Clee in 1980 (*SBR*) but there have been no subsequent records. A population maximum of 100 pairs seems likely.

The only more southerly locations for Red Grouse in England are Dartmoor and Exmoor, where they were definitely introduced, so safeguarding the small Shropshire population is a priority.   *TW*

RED-LEGGED PARTRIDGE

Alectoris rufa

Status: **Resident**

Tetrads with evidence of breeding

Confirmed	277 (32%)
Probable	253 (29%)
Possible	76 (9%)
Total	**606 (70%)**

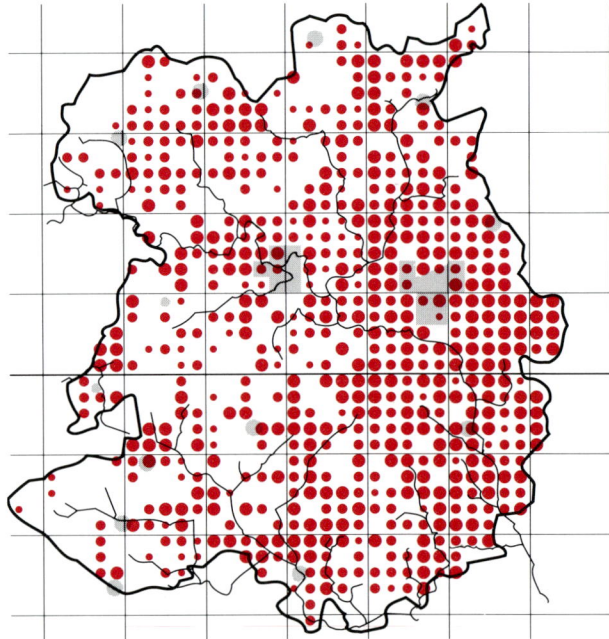

Red-legged Partridges were first successfully introduced in Suffolk in 1790, but the earliest date on record for Shropshire is 1877 (Forrest 1899), since when a sedentary feral population has become established, supplemented by releases, which latterly have been large and frequent. It is not known whether the earliest records were of releases or immigrants, but in describing later records, Forrest (1908) referred both to some sites where eggs were hatched under Grey Partridges and to others where birds "came of their own accord".

Forrest described the Red-leg as "rare" (1899) and "decidedly uncommon" (1908) but by the time of the *Handlist* (1964) it had become "locally fairly common". Fieldwork for the BTO *Atlas* (1976) did not confirm breeding in all 10-km squares, but this is now forthcoming, suggesting some consolidation of the population, probably the result of an increase in releases since 1974 (*Pop. Trends* 1990).

Red-legged Partridges are relatively easy to detect in spring, when crops are short and males are giving their loud "steam engine" calls. Nests are rarely found by chance but family parties may well be.

Both the BTO *Atlas* (1976) and *Wintering Atlas* (1986) show that Red-legs have a strongly southern and eastern British distribution, with Shropshire at the very north-west corner of the range. The east of the county is the stronghold, there being a marked thinning out of records to the north-west and west. The reasons for this "continental" distribution lie in climate and land use. Generally, areas where rainfall exceeds 87cm are spurned and arable farmland is preferred (*Pop. Trends* 1990). Comparisons with Grey Partridge distribution are made under that species.

In 1968 another *Alectoris* partridge, the introduced Chukar *A. chukar*, was first recorded in Britain. By 1972 both Chukars and Red-legged X Chukar hybrids were being released at many locations (Potts 1989) but it was not until 1986 that they were first reported in Shropshire (*SBR*). Very few Atlas records have distinguished between Red-legs, Chukars and hybrids and all records have been combined.

The BTO *Atlas* (1976) gave an average national CBC figure of 1.2 pairs per sq. km but it is thought that current feral populations of Red-legs in Shropshire rarely reach this figure. Even assuming four pairs per occupied tetrad, the population would only be in the order of 2000 pairs, which would doubtless fall if releases ceased.   *TW*

GREY PARTRIDGE
Perdix perdix

Status: **Resident**

Tetrads with evidence of breeding

Confirmed	269 (31%)
Probable	261 (30%)
Possible	93 (11%)
Total	**623 (72%)**

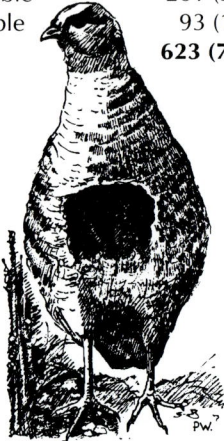

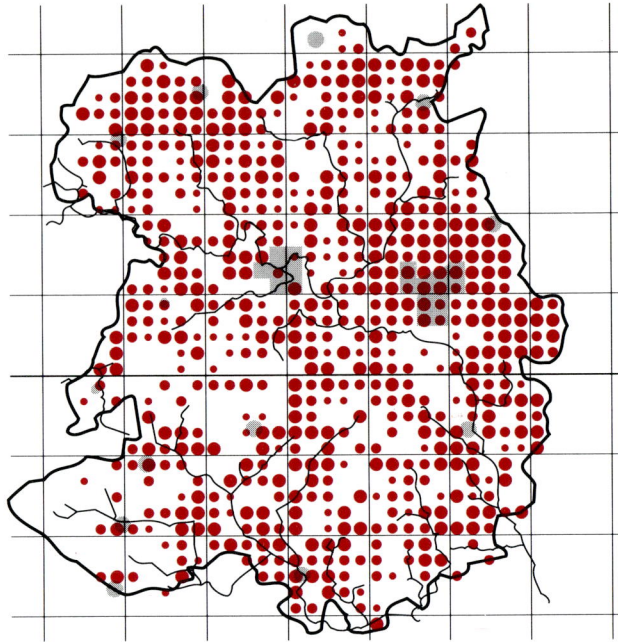

Cryptic coloration and crepuscular habits make Grey Partridges inconspicuous and they will have been overlooked in places. The best means of location is often the "rusty gate" skirl call. Many records are of pairs, which makes the recording of breeding, at least at the probable level, straightforward. Confirmation often comes from seeing family parties.

Comparison of this map with that for Red-legged Partridge is instructive. Both species thin out to the west and north, but the effect is less obvious in the case of Grey than Red-legged. Seemingly, Grey Partridges are generally more tolerant of higher rainfall and a higher proportion of grassland than the more "continental" and arable-orientated Red-legs (see Maps 3, 5 and 7), though the latter were reported "at home on higher ground at Brown Clee and Long Mynd" (*SBR* 1988) and "more tolerant of higher ground and wilder country than Grey Partridge" (*SBR* 1985).

At one time 2 million or more Grey Partridges were shot each year in Britain, but now the entire early autumn population averages less than half that number (BTO *Winter Atlas* 1986). Declines have occurred throughout the range and Shropshire has not been spared. Forrest (1899, 1908) described it as "very common" and the *Handlist* (1964) as "common", but fairly common might be a truer description today.

The reduction in the number of gamekeepers explains some of this marked decrease because nest predation is high and fox and corvid control contribute significantly to nesting success. Loss of hedgerow nesting habitat has also been a factor but more significant has been the development of herbicides to control agricultural weeds which harbour the insects on which Grey Partridge chicks depend (*Pop. Trends* 1990).

Subjective impressions gained during Atlas fieldwork suggest that one pair per sq. km in tetrads where they were recorded might be a current average, giving a population estimate of around 2500 pairs. Although releases do occur, they are infrequent nowadays and neither the distribution nor population level is thought to be materially affected.

TW

QUAIL
Coturnix coturnix

Status: **Summer visitor**

Tetrads with evidence of breeding

Confirmed	11 (1%)
Probable	65 (7%)
Possible	87 (10%)
Total	**163 (19%)**

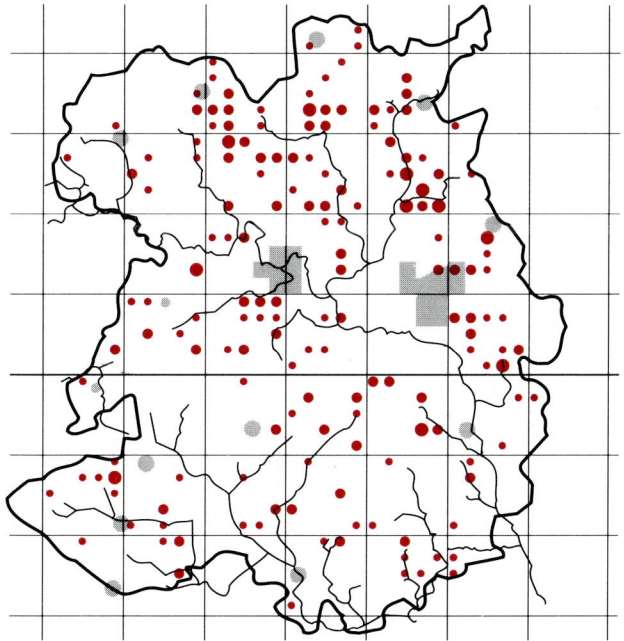

All records.

The Quail winters in Africa. Arrival is usually heralded by a far-carrying "whip-we-whip" call, particularly noticeable on warm evenings towards dusk. Whilst seeming to come from nearby, a well-hidden individual up to 1km away may be responsible. Fields of barley and to a lesser extent other cereals are preferred, with a minority found in long grass and even root crops.

Generally numbers reaching this country are quite small but irregularly much larger influxes occur. 1989 was a "Quail year" *par excellence,* when over 130 records were received compared with a total for the previous 32 years of only 119. Although the 1989 figure is comparatively inflated by reports received in response to newspaper appeals by the SOS and SWT, it still reflects what was an exceptional summer.

Two maps are published: one covers the total survey period whilst the second excludes 1989 to illustrate the more usual pattern. Both illustrate the scattered distribution. Rarely leaving dense vegetation, Quail are seen infrequently and the vast majority of records refer to calling males. Most sightings are made outside the fieldwork period by farmers harvesting in August. Calling is by no means an accurate guide to breeding density: unpaired males may move regularly over substantial distances in search of a mate, making report duplication quite likely, but calling is greatly reduced in less competitive areas once pairing has taken place

(*Pop. Trends* 1990). Regular calling may therefore indicate a breeding territory adjacent to other territories or an unpaired male seeking a mate. Even if calling does indicate an established territory, renowned ventriloquism in what is often dense vegetation confounds all but the most persistent or fortunate of observers when seeking proof of breeding.

In "non-Quail years" numbers still fluctuate, though at a much lower level. There has been a definite upward trend during the last decade (*SBR*s), but this may only be due to increased numbers of SOS members able to recognise the call. Some sites, such as Sleap/

All except 1989 records.

Noneley and Lilleshall, are used in most years and some reports refer to several adults, although few refer to more than one calling male. Most sites produce only isolated records, so the map excluding 1989 still overstates the usual position and a population of 5–8 pairs may be a reasonable estimate for most years.

As the much larger numbers in "Quail years" can be accommodated, shortage of habitat is unlikely to be a factor limiting the population. However, earlier cutting for silage has increased the likelihood of desertion or destruction of nests.   *DS*

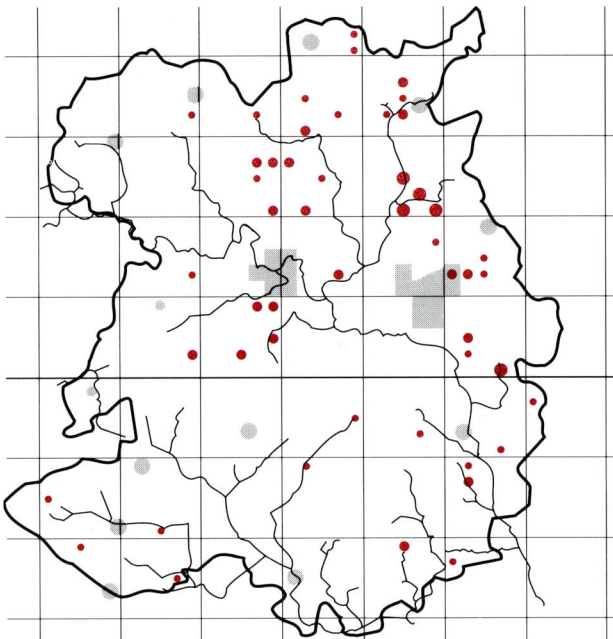

PHEASANT
Phasianus colchicus

Status: **Resident**

Tetrads with evidence of breeding

Confirmed	514 (59%)
Probable	198 (23%)
Possible	137 (16%)
Total	**849 (98%)**

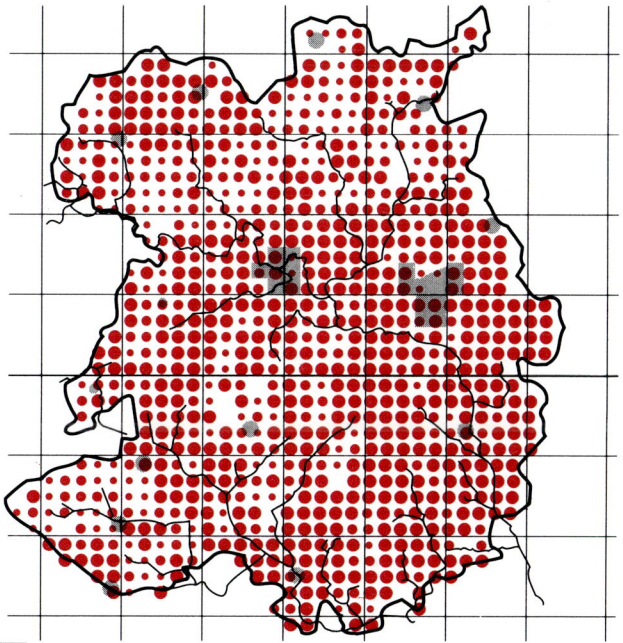

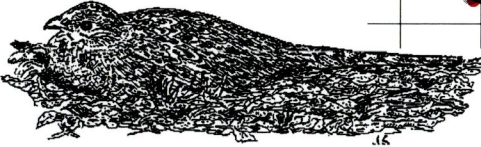

Anyone who has happened upon a Pheasant with young will long remember the cacophony of sound which results. The country lane "car obstacle", the whirring take-off or the "krok kok" call in the copse are more frequently encountered. It is resident and sedentary, and gradually formed a feral population after originally being introduced centuries ago.

Preferred habitats are copses, farmland with cover from long grass, and hedge bottoms, but they will nest in a wide range of sites, even in marshy areas on The Long Mynd and among nettles in a country garden! Proof of breeding may come from observing young or finding eggshells, and in many instances by asking local people. Breeding Pheasants are widespread, the gaps reflecting their absence from urban areas and high ground without copses.

Though obviously abundant, estimation of breeding numbers is complicated by two major factors. One is the release of large numbers of artificially reared Pheasants for the shoot; this in turn leads to them often being ignored as wild breeding birds, and a much larger population than could be sustained naturally. The other arises from the uneven effect of releases and shooting on the population balance and mating behaviour, which may result in large numbers of non-territorial males, male birds on territory but without a partner, a more equal pairing, or conversely, males with harems of up to ten females! A recent national study of population densities in the breeding season found an average per sq. km of 10 territorial males, 7 non-territorial males and 16 females (*Pop. Trends* 1990). Based on these figures the populations would be 36,000, 25,000 and 58,000 respectively, but obviously no estimate of breeding pairs is possible.   *DS*

WATER RAIL
Rallus aquaticus

Status: **Resident**

Tetrads with evidence of breeding

Confirmed	1	(0%)
Probable	4	(0%)
Possible	13	(1%)
Total	**18**	**(2%)**

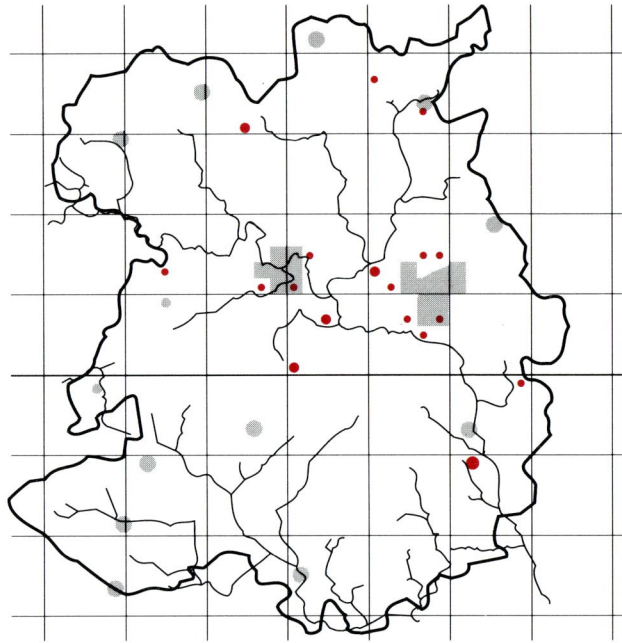

Water Rails rarely leave dense waterside vegetation. Winter produces the vast majority of records, when the resident population is supplemented by migrants from the Continent, the deep cover dies back somewhat, and hard weather often forces them into the open to feed.

Small numbers do breed, mainly in extensive reed beds on the margins of rivers, canals, lakes and reservoirs. Secretive behaviour and a tendency to be silent by day, together with inaccessible and decreasing habitat, ensures extremely few records. Finding fluffy juveniles by chance has provided some evidence of confirmed breeding.

Pride of place must go to Chelmarsh, where Water Rails have been seen to breed successfully in the marsh at the northern end of the reservoir, part of the SWT and SOS nature reserve, in four of the six years of Atlas fieldwork. They also probably bred at Astbury in the same tetrad in 1986. The four other probable breeding records came from Venus Pool, Allscott Sugar Factory and near English Frankton, each in only one year, and from Frodesley in both 1988 and 1989.

Because they are so elusive, little is known about population levels and trends. Residents are largely sedentary, so breeding probably continues at previously successful sites. Combining all breeding season records from the main Atlas map above, the *Handlist* (1964), the BTO *Atlas* (1976), and *SBR*s since then probably gives a fuller picture of the distribution of this scarce and overlooked species. Even over this 35-year period, there are only 6 confirmed, and up to 9 probable and 6 possible breeding sites. Pairs may be present at all these sites in most years, though numbers may perhaps be reduced following hard winters. Others are likely to be overlooked, implying a minimum of 20 breeding pairs, but there may be more.   *LS*

CORNCRAKE
Crex crex

Status: **Possible**

Tetrads with evidence of breeding

Confirmed	0 (	0%)
Probable	0 (	0%)
Possible	10 (	1%)
Total	**10 (**	**1%)**

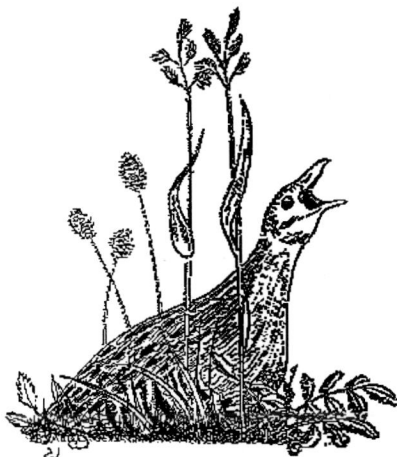

The Corncrake has not been known to breed here for 15 years, and is now only a scarce passage migrant.

Keeping to tall vegetation, but with a loud far-carrying call, it is far more often heard than seen. Where breeding does occur nests are usually in grass grown for hay, but other tall vegetation may be used.

Major long-term decline has continued for over a century (BTO *Atlas* 1976), but the recent acceleration has been dramatic. A survey in 1988 found just under 600 calling Corncrakes throughout Britain, only 5 of which were outside Scotland. Less hay production, a reduction in plant diversity and earlier cutting of silage on farms are the main causes (*Pop. Trends* 1990).

This drastic decline is mirrored here. Six years of Atlas fieldwork produced only 10 records, from widely scattered locations. All except two were of individuals, most of which were present for only one or two days. No records were received in 1987 or 1990. No map is published, as no evidence of even probable breeding was found, but the records are listed in *SBR*s.

The *Handlist* (1964) described it as "regular in small numbers, occasionally breeds", "widespread up to 1910" with records "much more scanty" after 1925. Four breeding records for the late 1950s, three confirmed and one probable, were noted. The BTO *Atlas* (1976) mapped six records, two for each of confirmed, probable and possible breeding. *SBR*s list a possible territory in 1983 and an unconfirmed report of a nest near The Wrekin Golf Course in 1984. The last records of confirmed breeding were from "an undisclosed site near Shrewsbury" (1972) and Northwood, near Wem (1975).

Unless there is a radical change in farming methods or land use, Corncrakes are unlikely to breed again in Shropshire.   *LS*

MOORHEN

Gallinula chloropus

Status: **Resident**

Tetrads with evidence of breeding

Confirmed	618 (71%)
Probable	37 (4%)
Possible	63 (7%)
Total	**718 (83%)**

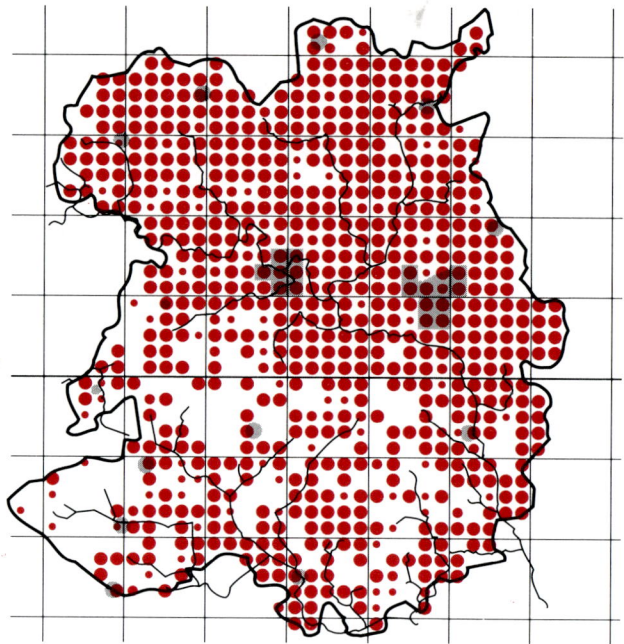

The Moorhen is one of the most characteristic birds of the north Shropshire plain, where the many hundreds of small meres and pools each support one or more pairs.

However, being adaptable, pairs are also found in a wide variety of other riverine and marshy habitats. Rivers, canals, sewage farms, flooded gravel pits, wet meadows, reed beds, willow and alder carr, and almost any stream, ditch or pond with surrounding cover will hold breeding Moorhens. This habitat is scarce in large areas to the south and west of the Severn valley, though the absence of Moorhens from the uplands of The Long Mynd, Stiperstones and Clun Forest is due to their dislike of fast-flowing streams rather than altitude *per se,* as testified by confirmed breeding on Boyne Water at 455m on Brown Clee.

Moorhens are easy to locate in late March and April when territorial activity and egg-laying begin. Males build platforms of water-weed for display purposes and more substantial structures are built for incubating eggs. Late nesting is common for repeat laying after the failure of first clutches, and sometimes for raising second or, rarely, third broods. During incubation, Moorhens are surprisingly easy to overlook, though their bulky nests can usually be found except in the densest cover. Once the chicks have hatched the brood is moved away from the nest, though the adults build temporary shelters for roosting.

Flooding is a major cause of nest failure for pairs selecting river or stream sites. Many waterways are subject to annual and sometimes very rapid flooding, especially in spring and often as late as May. Evidence from a WBS site on the River Tern suggests that many Moorhens avoid building nests there early in the season and only use the emergent vegetation in June and July when the risk of flooding is much less.

Though numbers are reduced by hard winters, especially true of 1962–63 (*Handlist* 1964), they recover quickly and nationally the population level is generally stable (*Pop. Trends* 1990). The average CBC farmland density of 3.8 pairs/sq. km

(BTO *Atlas* 1976) is likely to be exceeded in the excellent habitat of the northern plain, though Moorhens are much more scarce in the south and west. An average of 5–10 pairs/occupied tetrad suggests a population of around 3500–7000 pairs.

Watercourse management involving clearance of emergent vegetation reduces available habitat, though perhaps only temporarily. More permanent loss results from the continuing practice of in-filling farm ponds. The growing population of mink may, however, be the most serious threat to the Moorhen. The WBS site on the River Tern, recently colonised by this mammal, has suffered a 75% reduction in Moorhen numbers.

GT

COOT
Fulica atra

Status: **Resident**

Tetrads with evidence of breeding

Confirmed	276 (32%)
Probable	37 (4%)
Possible	26 (3%)
Total	**339 (39%)**

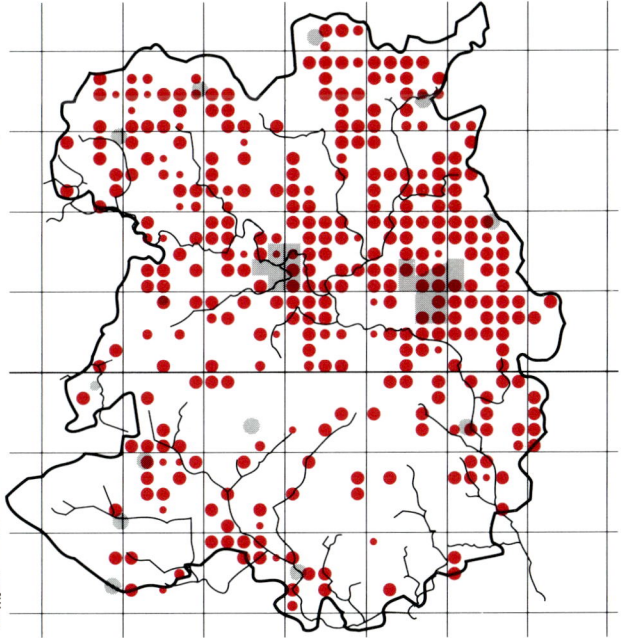

Once located on a pond or lake, the Coot is perhaps one of the easiest species to prove breeding, as reflected in the high ratio of confirmed records (81% of tetrads observed). Its aggressive nature towards individuals of its own and other species is especially obvious in April and May on shallow, eutrophic lakes which are the favoured habitat.

Some emergent or floating vegetation is required to build the often conspicuous nest, and breeding confirmation may also be obtained by watching family parties, often visible over long distances on open water.

Most breeding occurs in the north and east. It was also confirmed on Boyne Water at 455m on Brown Clee in the early Atlas period, though the emergent vegetation has now been cleared, so the distribution mirrors that of suitable lakes and pools rather than any particular altitudinal barrier. Larger pools are needed than those sufficient for Moorhens, hence the population is more restricted.

Though preferring still-water sites, breeding also takes place on slower stretches of the Severn at Atcham, Melverley and Shrawardine, and on the lower reaches of the Tern (*SBRs*).

Where the Coot does occur, associations of 10 or more pairs are usual on the larger pools, with only isolated pairs on small farm ponds or rivers. The recent increase in man-made still-water lakes such as reservoirs and gravel pits has enabled the population to expand (*Pop. Trends* 1990). An average of 3–6 pairs per occupied tetrad suggests a population of at least 1000 pairs but probably no more than 2000.

GT

OYSTERCATCHER
Haematopus ostralegus

Status: **Summer visitor**

Tetrads with evidence of breeding

Confirmed	7 (1%)
Probable	2 (0%)
Possible	1 (0%)
Total	**10 (1%)**

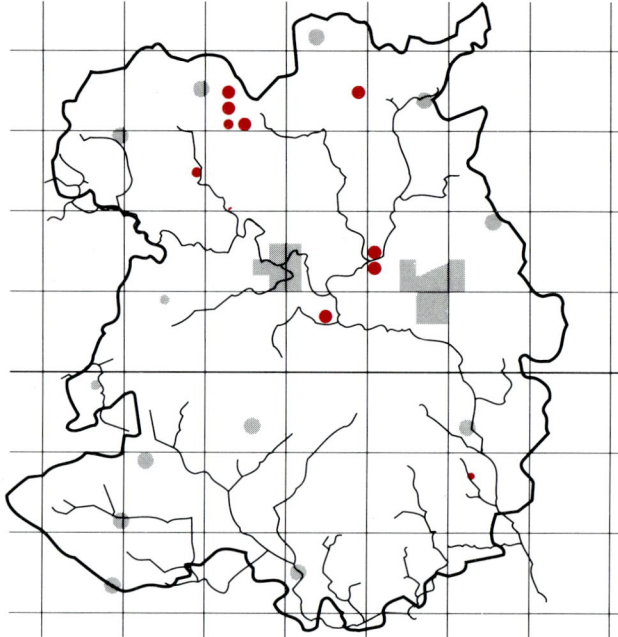

The Oystercatcher is unmistakable, but still seems strangely out of place to those used to seeing it in the more familiar habitat of the seashore. The *Handlist* (1964) recorded it as a scarce passage migrant and there were no records during fieldwork for the BTO *Atlas* (1976). By the end of the 1970s the number of *SBR* records was increasing and the first confirmed breeding occurred near Lyneal Wood in 1981.

The cluster of dots near Ellesmere is associated with the meres and nearby wet areas. Breeding is now a regular occurrence at Allscott and near Isombridge, and a pair laid eggs at the SOS Venus Pool Reserve in 1990, although no young were raised.

Habitats occupied include gravel pits and arable farmland within a reasonable distance of damp feeding areas. Generally the large size, striking plumage, characteristic far-carrying call, and, when they are well-grown, the conspicuous young, ensure pairs are located, though a few were reported by farmers, enabling breeding to be confirmed where it might otherwise have been overlooked.

Inland breeding is commonplace in northern England and Scotland and, as the population grows, the habit appears to be spreading south (*Pop. Trends* 1990), so the Atlas documents the beginning of the trend in Shropshire. Numbers will almost certainly increase from the the present six or more pairs as there are many other sites that offer similar suitable habitat.

CEW

LITTLE RINGED PLOVER

Charadrius dubius

Status: **Summer visitor**

Tetrads with evidence of breeding

Confirmed	7 (1%)
Probable	2 (0%)
Possible	8 (1%)
Total	**17 (2%)**

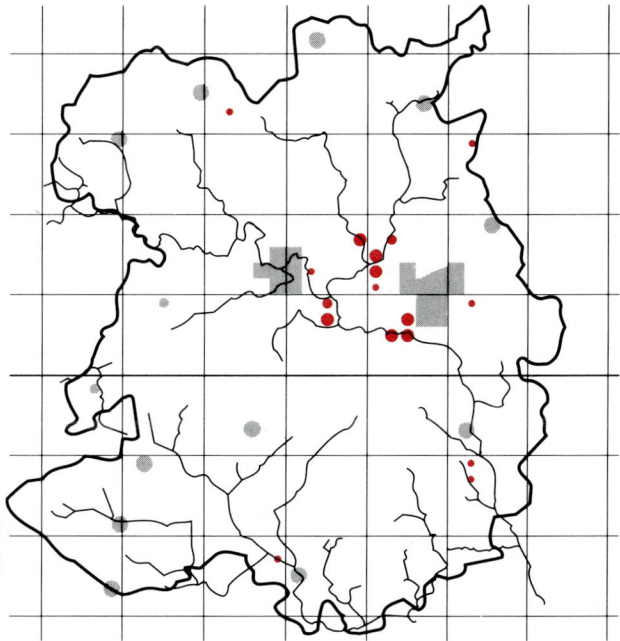

The Little Ringed Plover was not known to breed in Britain until 1938. First recorded in Shropshire in 1957, the *Handlist* (1964) described it as a "passage migrant, scarce". Breeding was not proved until 1976 and by 1984 it was being reported from three sites but suspected at others (Sankey *SBR* 1984). Still a scarce summer visitor, a few pairs breed each year.

They nest on the bare margins of wetlands. Sand and gravel pits are favoured in Britain and are used here, including the SOS Venus Pool Reserve, which was originally such a site and is now actively managed to attract them. Other habitats include bare or spring-sown arable fields close to wetlands, and shingle bars in rivers, although fluctuating water levels and regular trampling by cattle and humans can reduce breeding success there, especially along the Severn. However, the plovers are tolerant of disturbance and lay repeat clutches if the first is destroyed. Some of the sites are temporary, for example the Swan Farm open-cast coal mine is being reclaimed and no longer has suitable habitat. Venus Pool and Allscott regularly hold 2–4 pairs each.

Little Ringed Plovers can be located through their very specific habitat requirements and are conspicuous during display. Once they have eggs or young, the adults even draw attention to themselves by landing in front of an intruder, putting on a spectacular injury-feigning distraction display. The chicks may be visible from some distance at the more open sites. Breeding is, therefore, fairly easy to confirm. The small dots may well refer to migrants although suitable breeding areas do exist in each of these tetrads.

They have not been reported from a number of sand and gravel pits where access is usually restricted, and some of the industrial sites are unlikely to be popular with observers so a few more pairs may be present. The national population is still increasing, especially in the west and north, with over 600 pairs in 1984 (*Pop. Trends* 1990). As Shropshire is at the edge of the British range, the present population of about 10 pairs is expected to increase.   *CEW*

LAPWING
Vanellus vanellus

Status: **Resident**

Tetrads with evidence of breeding

Confirmed	411 (47%)
Probable	249 (29%)
Possible	83 (10%)
Total	**743 (85%)**

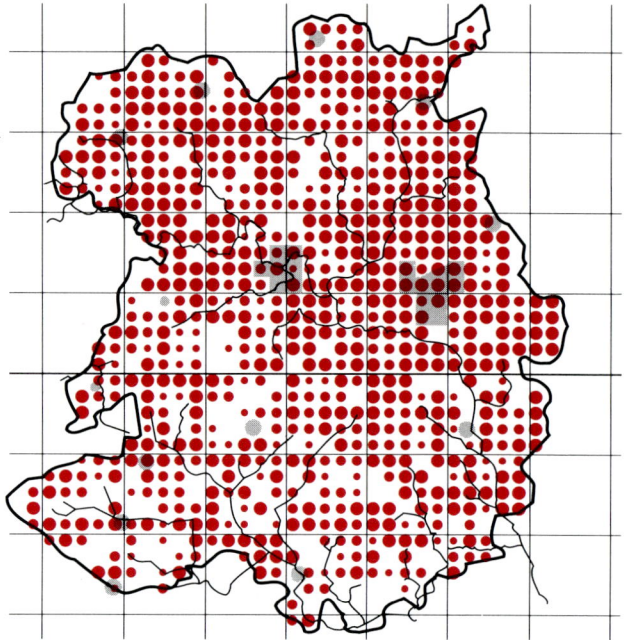

A rather thinly distributed resident, the Lapwing is also a winter visitor in some numbers, especially to the lowlands. The tumbling display flight and characteristic "peewit" call of the male are difficult to overlook. The Lapwing prefers a mosaic of arable and pasture land, nesting in the relatively bare fields of the spring sowings and then moving the young to the adjacent grassland. It is also found on upland grasslands and on the edge of wetlands, including gravel pits and sewage farms. The nest and chicks can be easily located in this open habitat. Upland moorland, woodland, uniformly arable areas and towns are unsuitable, though it does breed on damp derelict sites within Telford.

The *Handlist* (1964) recorded the Lapwing as a common resident, but some observers noted a decrease since the 1930s and evidence of a further decrease after the hard winter of 1962–63. Now it is not so common and the national trend is downwards, especially in central and southern England, due to changes in farming practice (*Pop. Trends* 1990).

Similarly, though much of Shropshire has previously provided the variety of habitat necessary, the strong tendency towards autumn sowing has reduced the available bare ground in spring in cereal fields. Grasslands are managed to produce a rich growth which is cut early for silage and is less suitable for the small chicks. It is likely that insufficient young are now being raised to maintain the population in lowland Shropshire, although more research needs to be done to be sure.

In 1987 a survey of 34 randomly selected tetrads (one from each 10-km square) located 119 nesting pairs, of which 87 were on arable and 32 on grassland. Of the 87 pairs on arable 40 were on bare land, 18 in spring cereals, 17 in autumn cereals and 12 in other crops. Of the 32 on grassland 20 were on ungrazed sites and therefore at risk if silage was cut. Pairs per tetrad varied from 0 to 22, with an average of 3.5 (unpublished BTO data for Shropshire). Appying this average to each occupied tetrad gives a population of 2300 pairs, probably on the low side in 1987, but the continuing decline suggests that this is now almost certainly too high.   *CEW*

SNIPE
Gallinago gallinago

Status: **Resident**

Tetrads with evidence of breeding

Confirmed	11 (	1%)
Probable	47 (	5%)
Possible	70 (	8%)
Total	**128 (**	**15%)**

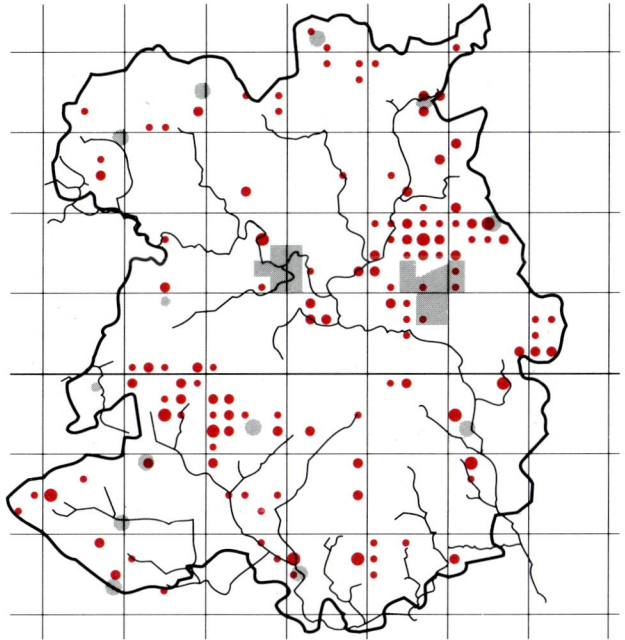

The Snipe is best known as a winter visitor and passage migrant in considerable numbers, but some also breed. The characteristic "drumming" display is unlikely to be overlooked if the observer is in the right place at the right time, and has given rise to many of the probable breeding records. Apart from this display the Snipe is secretive and elusive, and the habitat is difficult to work, so finding the nest or chicks to confirm breeding is unusual and the map will under-represent the distribution. A few of the small dots may relate to late passage birds.

The Snipe occurs in two distinct breeding habitats, the wet hollows of the uplands and the damp meadows of the lowlands (Map 5). Most of the records in the north and east are from river valleys, with a marked concentration in the now much-drained Weald Moors north of Telford. The other concentration in the south-west marks the uplands of the Stiperstones and The Long Mynd. Elsewhere the distribution is patchy, suggesting isolated areas of suitable habitat.

The *Handlist* (1964) recorded that the Snipe was "found fairly commonly throughout the county" but noted a decline as a result of intensive land drainage. Less common now, breeding has not been confirmed in three 10-km squares in the north-east where it was proved for the BTO *Atlas* (1976). Numbers are still declining rapidly on the lowland meadows (*Pop. Trends* 1990) and this is attributed to continuing drainage. More intensive grazing is an added pressure as nests are more likely to be trampled. Many meadows have been converted to arable use, reducing still further the area of grassland suitable for breeding.

The BTO *Atlas* (1976) estimated 30–40 pairs per occupied 10-km square, but, in view of the considerable decline since, this would probably overestimate the current Shropshire population. With only 57 occupied tetrads, many of them isolated, even after assuming that most of the small dots represent breeding pairs and there is some under-recording, it is unlikely that the breeding population exceeds 200–300 pairs.

CEW

WOODCOCK
Scolopas rusticola

Status: **Resident**

Tetrads with evidence of breeding

Confirmed	36 (4%)
Probable	72 (8%)
Possible	51 (6%)
Total	**159 (18%)**

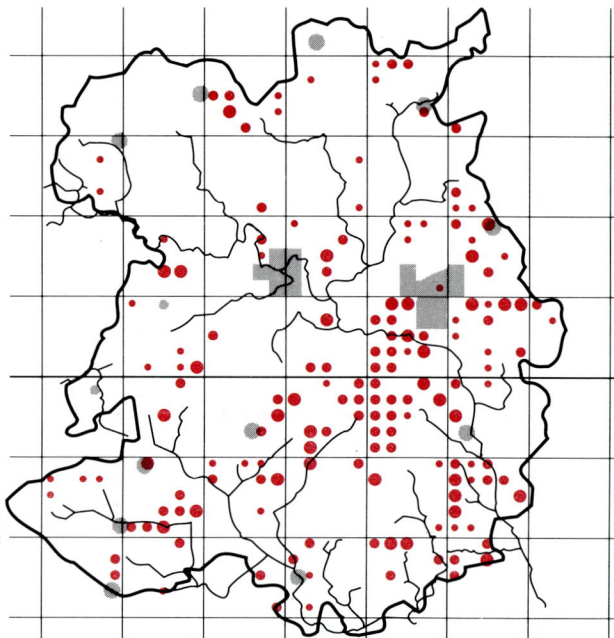

A resident and winter visitor, the Woodcock mainly inhabits open deciduous woodland with dry ground for nesting and damp areas to feed. It is also found in mixed woods and young conifer plantations. The *Handlist* (1964) stated it was nowhere common but recorded from woodlands throughout the county. Atlas fieldwork has confirmed this fact and the distribution shown correlates closely with woodland (Map 8). The Wenlock Edge woodlands and Haughmond Hill are noted strongholds (*SBRs*).

A special effort is required to locate the Woodcock, which is rarely seen by day unless disturbed. Many of the records refer to the evening display flight, or "roding", of the male as he patrols the woodland seeking a mate. As evening visits to many sites were not undertaken it is almost certainly substantially under-recorded, and the number of records is encouraging. A roding male does not defend a territory and may visit several woods in the search for a mate. Once the eggs are laid he will resume roding in the search for another female, leaving the first to raise her young. Effective camouflage in extensive woodland makes finding nests or downy young very difficult, so less than a quarter of the records are of confirmed breeding. Even the smallest dots probably indicate breeding, although roding males may visit more than one tetrad. Assuming one or two nests in each tetrad recorded, a very rough estimate of 150 to 300 "pairs" is obtained though under-recording may render this too low.

Nationally some decline is suspected in the south due to loss of woodland habitat (*Pop. Trends* 1990) but there is insufficient information available to assess local trends. Absent from ten 10-km squares in the BTO *Atlas* (1976), it is now only absent from five, but this may only reflect more fieldwork over a longer period. Much ancient woodland has been lost since 1900, continuing until around 1985, reducing the habitat available. In 1979, 483 woodland sites covering 6019 ha were designated as Prime Sites for Nature Conservation by the SWT, but within 10 years 9% of these had been lost to agriculture, forestry or development (SWT 1989). Proposals to establish a forest belt along the Severn valley from Telford to the Wyre Forest cannot restore these ancient woodlands but may, in time, go some way to redress the balance.

CEW

CURLEW

Numenius arquata

Status: **Resident**

Tetrads with evidence of breeding

Confirmed	184 (21%)
Probable	322 (37%)
Possible	155 (18%)
Total	**661 (76%)**

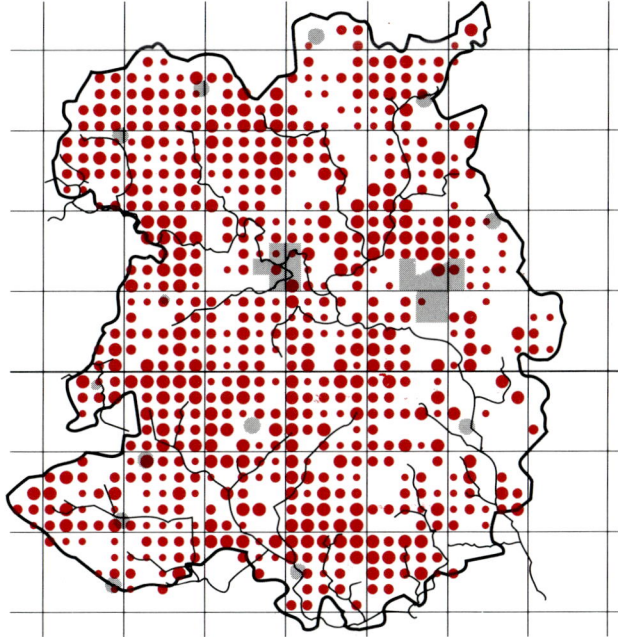

JS

Present throughout the year, the breeding population is supplemented by passage and winter visitors from the Continent. The *Handlist* (1964) recorded that the Curlew was "common, found in suitable habitats throughout the county" and described the expansion of the breeding range from the hill country to the lowlands which began in 1913. This was confirmed by the BTO *Atlas* (1976) which recorded Curlew in all but one eastern 10-km square.

The pattern is now repeated, with some large gaps in the east which appear to correlate with Telford, and areas of cereal crops (Map 7) and minimum rainfall (Map 3). Drainage of damp pastures and the change from haymaking to silage have reduced the area of optimum habitat available, but some still remains and its conservation would benefit other species as well as Curlew. The damp meadows of the Severn/Vyrnwy confluence, The Weald Moors and Tern valley are particular lowland strongholds. Though widespread in the uplands, there are some unexplained gaps in the Clun Forest area which deserve further study.

Early-season visits will locate displaying birds which can be heard from a distance. After egg-laying they are much quieter and could then be overlooked. Probable breeding is easy to establish and most of the small dots reflect tetrads not revisited at the right time to observe territorial behaviour, though a few may relate to late passage birds. Confirmation of breeding is much more difficult, often relying upon the appearance of the young once well-grown. The BTO *Atlas* (1976) estimated 15–25 pairs per 10-km square and Shropshire is probably nearer the upper limit. Most tetrads in the uplands will have more than one pair but in the lowlands, away from the main concentrations, a pair may range over more than one tetrad and some may not be occupied every year. Taking an average of 1 pair per tetrad gives a population of about 700 pairs.

CEW

REDSHANK

Tringa totanus

Status: **Summer visitor**

Tetrads with evidence of breeding

Confirmed	8 (1%)
Probable	10 (1%)
Possible	14 (2%)
Total	**32 (4%)**

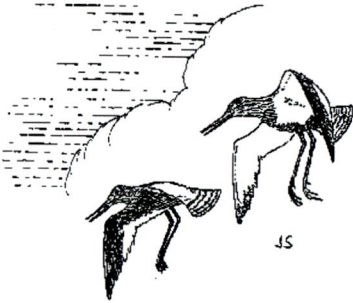

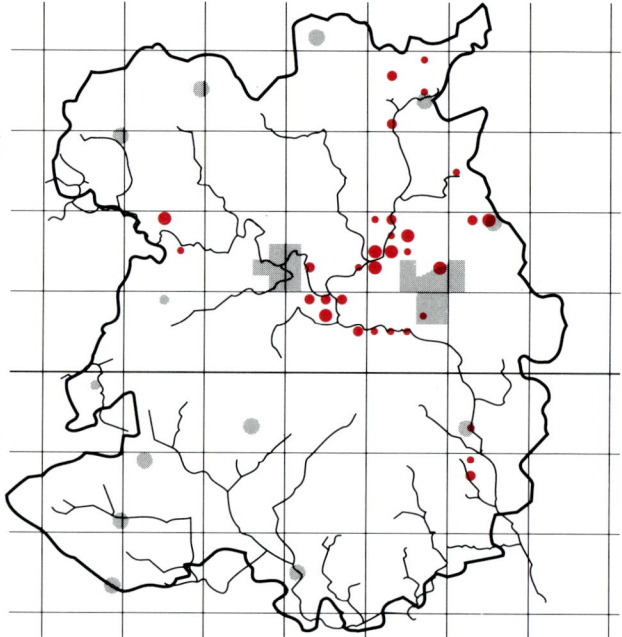

The Redshank is a passage migrant and scarce breeding visitor, rare in winter. Dependent on wetlands, it is now threatened by habitat loss due to the draining and ploughing of wet meadows.

The *Handlist* (1964) stated: "At the turn of the century [the Redshank] was considered an uncommon bird. It first bred in the county in 1910 at Crudgington and soon became well established there." It then spread along the Tern valley and reached several other localities by the 1930s. After noting a decline since 1940, probably due to extensive drainage, the *Handlist* (1964) continued that it still bred in fair numbers where suitable habitats persisted, especially on the floodplains of the Severn, Tern, Roden and Perry, on The Weald Moors and at Venus Pool. The BTO *Atlas* (1976) revealed a more restricted distribution, and a further contraction has since occurred in line with the national trend, which is strongly downwards, particularly in the inland counties (*Pop. Trends* 1990).

The river valleys still hold a few pairs, but with the virtual disappearance of wet meadows, the major habitat has become the damp margins of sewage farms, settling ponds and gravel pits. With such small numbers, the fragmentation of the habitat, and the use of quite small sites, some pairs were probably overlooked. In less well-covered tetrads the small dots may equally represent either breeding or passage birds.

Having located a pair, courtship displays and a noisy defence of the nest site indicate breeding but in many cases this is only confirmed when the chicks appear. They are easier to observe at sites with permanent hides such as Venus Pool Reserve. Elsewhere chicks may be missed as they are well camouflaged and freeze while agitated parents seek to drive off intruders, so most of the probable records will relate to breeding pairs.

It seems unlikely that there are more than 30 to 50 breeding pairs left. Whether the new financial incentives for farmers to maintain habitats such as damp riverside grasslands have arrived in time to prevent further population loss remains to be seen.

CEW

COMMON SANDPIPER
Actitis hypoleucos

Status: **Summer visitor**

Tetrads with evidence of breeding

Confirmed	11 (1%)
Probable	21 (2%)
Possible	37 (4%)
Total	**69 (8%)**

JS

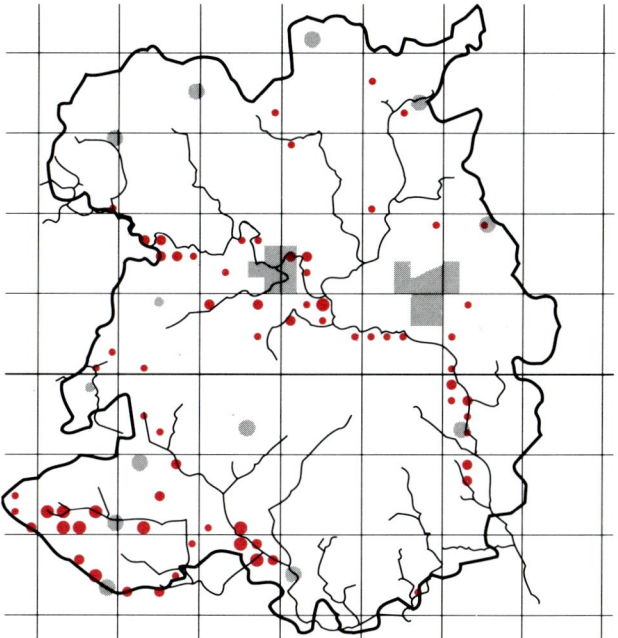

The Common Sandpiper is at the eastern edge of its main breeding range in this part of Britain and many migrate along the river systems. Most, if not all, of the possible and probable records from the Severn valley and the north will in fact relate to the singles and pairs which pass through as late as mid-May and return again from late June (*SBR*s). There are many other breeding season *SBR* records which clearly relate to migrants and are not included on the map.

Some pairs attempt to breed in the Severn valley using the shingle bars and stony shores that are exposed in the summer months but, in addition to nest loss from flash floods caused by heavy rain, they are subjected to considerable disturbance by fishermen and other river users, and to trampling by cattle. Young were successfully raised at the Severn/Tern confluence in 1987, but three eggs disappeared from a nest in the same place in 1988 (*SBR*s).

In the more remote south-west there is much better breeding habitat on the rocky shores of the faster-flowing streams, and here the rivers Teme, Clun and Onny and their tributaries have provided several confirmed breeding records, with five pairs in one season on about 20 km of the Teme between Bucknell and Felindre (*SBR* 1987). Adults are fairly conspicuous, but chicks are well camouflaged, so most of the probable breeding records in the south-west are likely to relate to breeding pairs.

Nationally the population is stable (*Pop. Trends* 1990). Records have come from more 10-km squares compared to the BTO *Atlas* (1976), probably due to better coverage over a longer period of fieldwork.

It is likely that only 1–2 pairs breed annually in the Severn valley but if most of the 18 occupied tetrads in the south-west support 1–2 pairs the population will be between 20 and 40 pairs. *CEW*

BLACK-HEADED GULL
Larus ridibundus

Status: **Resident**

Tetrads with evidence of breeding

Confirmed	8 (1%)
Probable	14 (2%)
Possible	35 (4%)
Total	**57 (7%)**

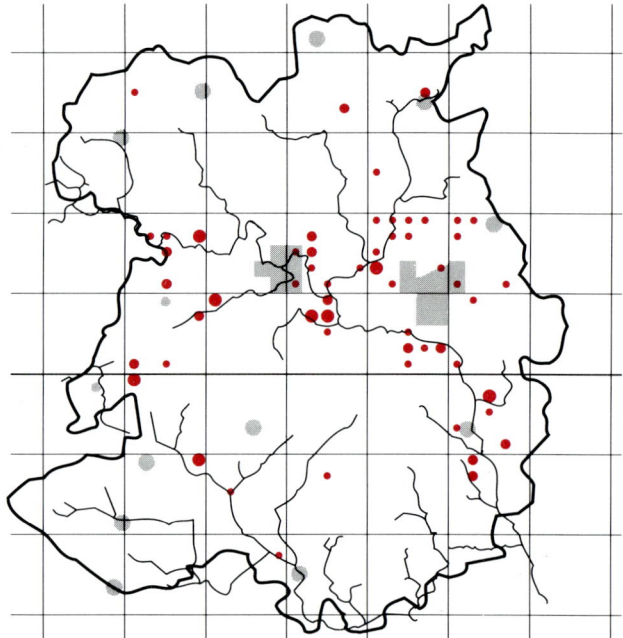

Present in large numbers during the winter months, only a few hundred pairs of this highly adaptable gull linger to breed.

Selecting sites on islands, or in low bushes or marshy vegetation surrounded by water, Black-headed Gulls are highly social but opportunistic breeders. They have the ability to quickly colonise newly created sites such as gravel pits and flooded meadows, and abandon them just as quickly if conditions become unsuitable.

The majority of occupied tetrads hold 1–6 pairs attempting to breed, often well away from the main colonies and usually unsuccessfully, for example at a small pool on Stapeley Hill at 310m.

The major colonies are at Cranmere Bog, Worfield, a traditional site occupied since at least 1958, and Venus Pool, but numbers vary dramatically from year to year (*SBRs*). The colony at Allscott is apparently now deserted with no nesting since

1988. Evidence suggests a great deal of interchange between the major colonies. High numbers at one site coincide with few if any breeding pairs at the other. Ringing recoveries testify to mobility between sites, as nestlings ringed at Allscott have subsequently been found breeding at both Venus Pool and Cranmere Bog, and an injured adult at Venus Pool in 1986 had hatched in Lincolnshire six years earlier. Recently 400 birds deserted Cranmere Bog in early spring, and Venus Pool appears to be the most favoured site in recent years.

Human disturbance is thought to have been a factor in the desertion of Cranmere Bog but does not appear to explain the abandonment of Allscott. Water levels are also thought to be important; not so high that nest sites become inundated, but high enough to act as an effective barrier to predation, especially by foxes, which were thought to have been responsible for the low breeding success at Venus Pool in the dry summer of 1989.

Surprisingly, Black-headed Gulls almost became extinct in Britain at the end of the last century, and were first noted wintering in Shropshire in 1939. Numbers increased, with unconfirmed reports of nesting on The Long Mynd in 1942, and five colonies were known in the 1950s (*Handlist* 1964). Three of these sites: Stockton Wood Farm, near Chirbury, where up to 125 pairs nested in 1950–52; Acton Pool, Clun; and most recently Allscott, are no longer used.

A survey in 1973 found three colonies with 386–436 nesting pairs, a substantial increase from 70–76 pairs at four colonies in 1958. However, one of these, at Brick Kiln Farm, Whitchurch, was a new and temporary colony with 86 nests in a flooded field (*SBR* 1973).

Now 100–200 nesting pairs would seem to be the extent of the breeding population.   *GT*

FERAL PIGEON
Columba livia

Status: **Resident**

Tetrads with evidence of breeding

Confirmed	88 (10%)
Probable	152 (17%)
Possible	162 (19%)
Total	**402 (46%)**

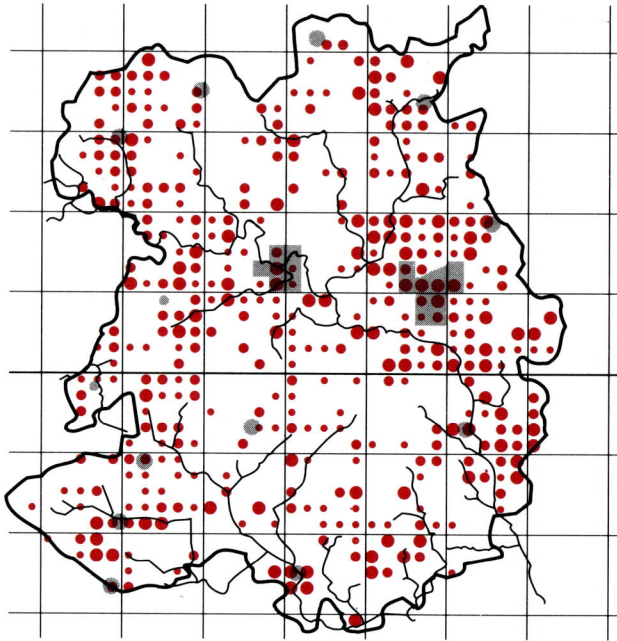

This descendant of the domestic pigeon and wild Rock Dove has become a well-established member of the British avifauna, especially in urban and farmyard habitats. However, ornithologists tend to ignore the Feral Pigeon; it was not mentioned in the *Handlist* (1964) and the County Bird Recorder still receives few reports, so the Atlas gives the first detailed assessment of its status since the BTO *Atlas* (1976).

Shropshire has few big towns, although Shrewsbury holds a large Feral Pigeon population, the true size and status of which is unknown. Hence it mainly occupies rural habitats, and is closely associated with farmsteads, barns and grain storage depots. The quarry cliffs of Wenlock Edge, a habitat akin to the coastal cliffs used by wild Rock Doves, have also been recorded as breeding sites (*SBRs*).

Domestic dovecotes and flocks of racing pigeons are common and confuse the overall picture, as non-returning individuals no doubt contribute to the continuing growth of feral flocks, and perhaps account for the high proportion of possible breeding records.

Urban populations of Feral Pigeons breed throughout the year (*Pop. Trends* 1990), but there is no information on whether the breeding season of Shropshire pigeons extends beyond that covered by Atlas fieldwork, though this may explain the low proportion of confirmed breeding records.

The distinct easterly bias in distribution confirms the association with arable farms. It is now present in all 10-km squares, which clearly reflects an increase since the BTO *Atlas* (1976), when 21 of the 10-km squares currently occupied had no record of breeding Feral Pigeon. Though undoubtedly under-recorded, almost by force of habit, it has now been noted in over 400 tetrads which, assuming 5–15 pairs each, suggests a population in the order of 2000–6000 pairs. *GT*

STOCK DOVE
Columba oenas

Status: **Resident**

Tetrads with evidence of breeding

Confirmed	271 (31%)
Probable	390 (45%)
Possible	106 (12%)
Total	**767 (88%)**

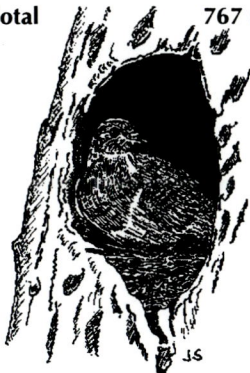

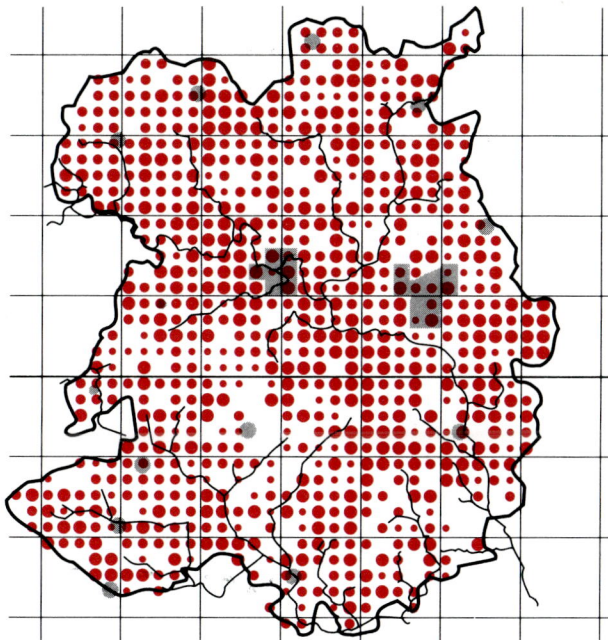

Although Stock Doves occasionally use large open nests such as those abandoned by Carrion Crows, or even squirrel dreys, they much prefer holes and crevices. Most nest in hollow trunks and boughs in the still-substantial standing of old deciduous timber. However, where stone quarries occur, such as along Wenlock Edge, Stock Doves are quick to take up residence so long as disturbance is limited. Abandoned farm buildings are also a favourite nest site.

Two or three broods per year are usual but Stock Doves do not have the protracted breeding season typical of the closely related Woodpigeon, Feral Pigeon and Collared Dove.

Their hole-nesting habit makes breeding difficult to confirm, and most such records are of recently fledged juveniles, or young in the nest hole heard calling. Conversely, probable breeding is easy to establish by virtue of the distinctive far-carrying display song and amorous behaviour in spring.

Stock Doves inhabit parkland and farmland, and, although the distribution shows no particular preference for lowland regions, gaps do occur in the highest hills, town centres and extensive woodland.

The *Handlist* (1964) described them as "resident in small numbers" but apparently much reduced since the turn of the century. This reduction was due to a widespread population crash in the 1950s and early 1960s, attributed to poisoning by organochlorine seed dressings. Subsequent recovery has not attained pre-1950 population levels, perhaps due to increased application of herbicides and the ploughing of winter stubbles (*Pop. Trends* 1990).

The CBC index suggests a twofold population increase since the BTO *Atlas* (1976) estimate equivalent to 2–4 pairs per tetrad nationally (*Pop. Trends* 1990). As much of Shropshire is good habitat, around 10 pairs per tetrad here is more likely now, suggesting a population of 7000–10,000 pairs. *GT*

WOODPIGEON
Columba palumbus

Status: **Resident**

Tetrads with evidence of breeding

Confirmed	742 (	85%)
Probable	119 (	14%)
Possible	5 (	1%)
Total	**866 (**	**100%)**

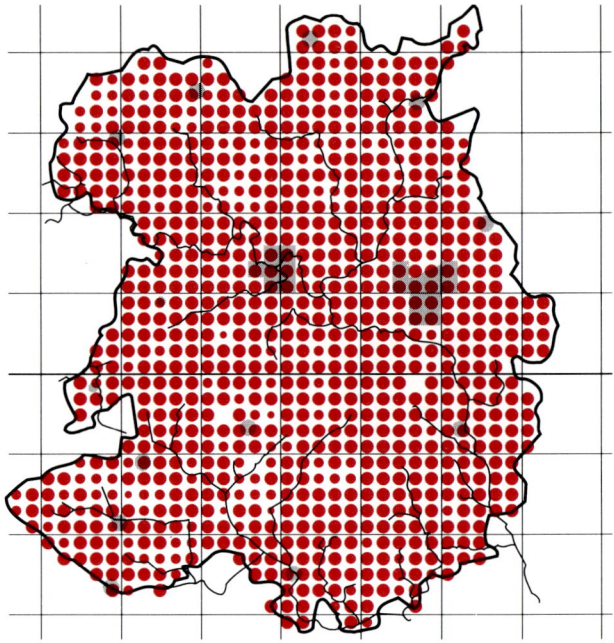

Abundant and classified as a pest, the Woodpigeon is very successful in Britain and Shropshire is no exception.

Noisy courtship behaviour, conspicuous nest and multiple broods make breeding easy to confirm, and nesting is believed to occur in every tetrad. The adaptability of its breeding cycle may explain this success, at least in part, as Woodpigeons have an extended breeding season, with eggs or chicks having been recorded from May until October (*SBRs*). Late breeding ensures access to newly harvested grain or natural seeds and fruits when the squabs are in the nest. The recent increase in autumn-sown cereals, which are harvested sooner, has resulted in a switch to early nesting in arable areas (*Pop. Trends* 1990).

Cereal production, though on the increase here, is far less common than in many English counties, so Shropshire Woodpigeons, especially those in the south-west, must be much more dependent on natural foods such as hedgerow weeds and woodland fruits. Even in the arable areas the population density depends on the crop regime, which affects the availability of winter food as well (*Pop. Trends* 1990).

An average equivalent to 40 pairs per tetrad (BTO *Atlas* 1976), with more in the cereal-growing areas, suggests a population exceeding 35,000 pairs.   *GT*

COLLARED DOVE
Streptopelia decaocto

Status: **Resident**

Tetrads with evidence of breeding

Confirmed	430 (49%)
Probable	281 (32%)
Possible	67 (8%)
Total	**778 (89%)**

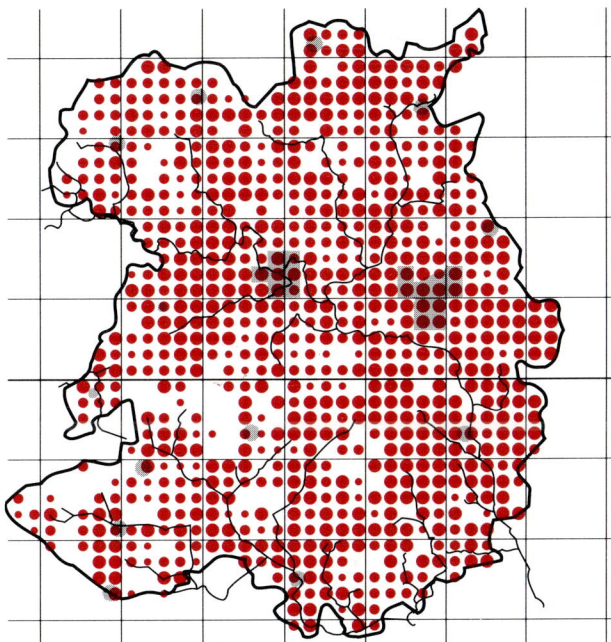

The remarkable colonisation of Britain in the late 1950s and early 1960s by the Collared Dove, after breeding was first proved in Norfolk in 1955, has been widely documented (*Pop. Trends* 1990).

In Shropshire it was first recorded on the Cronkhill Estate at Atcham in 1961 when a pair were seen, with two pairs there the following year. Breeding was first confirmed in 1963 at Ludlow (*Handlist* 1964) and by 1972 the whole county had been colonised (BTO *Atlas* 1976). The population continued to grow during the 1970s (*SBR*s), since when numbers appear to have stabilised.

As with other members of the pigeon family the Collared Dove is easy to find, particularly in the first four months of the year, as a result of its noisy and vigorous displays and close association with human habitation, especially villages, farms and suburban gardens. The flimsy nests may be in the most obvious of sites, commonly on ledges in barns, though evergreen trees and shrubs such as holly are usually selected. The young accompany the adults for several weeks after leaving the nest so breeding is, on the whole, easy to confirm.

Egg-laying has been recorded as early as January, with chicks still in the nest as late as October (*SBR*s), so the protracted breeding season allows up to 4–6 broods per year. As with the Woodpigeon, spilt grain has been a major food source in the past and has probably been instrumental in allowing the rapid spread throughout Britain. Though cattle farms use grain for winter feed, it is not given to sheep, perhaps accounting for the absence from predominantly upland areas.

Robertson (1990) speculated that the apparent stability of the population now may be due to control by humans. However, the more efficient harvesting and storage of grain crops, coupled with saturation of suitable territory, seem more plausible explanations. A mean density of 100 pairs per 10-km square or around 4 pairs per occupied tetrad (*Pop. Trends* 1990) indicates a population of 2000–3000 pairs. Hence it is now common and widespread, though with a tendency to avoid the bleaker upland regions.

GT

TURTLE DOVE
Streptopelia turtur

Status: **Summer visitor**

Tetrads with evidence of breeding

Confirmed	25 (3%)
Probable	100 (11%)
Possible	106 (12%)
Total	**231 (27%)**

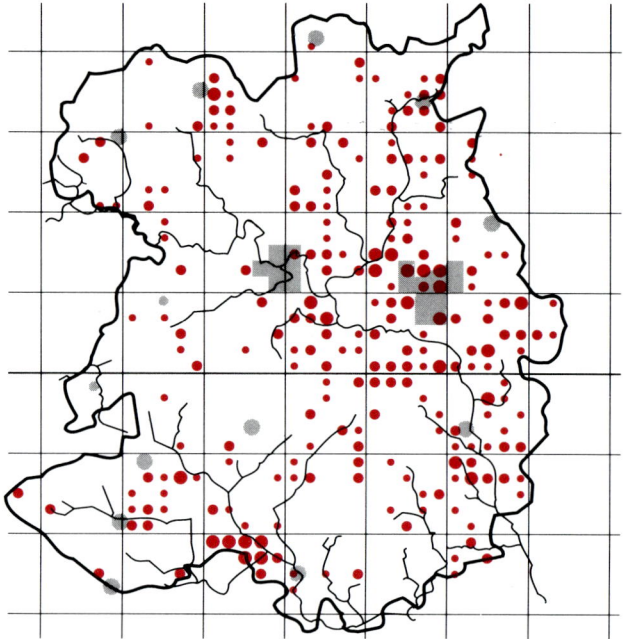

Turtle Doves usually arrive from early May but new immigrants are located into June. Two eggs are normal and two broods are usually reared before the return migration. Departure is very early, most apparently leaving in the second half of August (*SBRs*).

They require trees or bushes to build nests and open ground to feed, with arable farmland a favoured habitat. The distinctive call makes location easy but confirmation of breeding is difficult because finding the nest, or distinguishing young from the very similar adults, is necessary.

The western uplands are unattractive to Turtle Doves, and their distribution is distinctly skewed to the east. Pockets are found in the river valleys of the south-west and around the meres near Ellesmere, but they are more frequent east of Shrewsbury.

Comparison with the BTO *Atlas* (1976) suggests a recent withdrawal from the north-west consistent with a severe decline since the late 1970s throughout Britain (*Pop. Trends* 1990). This confirms anecdotal evidence from *SBRs* in recent years.

The seeds of arable weeds, especially those of common fumitory, are a major diet component on agricultural land in East Anglia (Murton *et al.* 1964). *The Flora* (1985) shows fumitory is widespread with no apparent trend towards an easterly distribution, so the Turtle Dove's range does not seem to be restricted by supplies of this proven food. However, the amount of food has fallen recently: the almost universal use of herbicides and fertilisers on grass leys and permanent grass, and the switch to early silage making, have resulted in less plant diversity, and hence a lower availability of weed seeds. Additionally Turtle Doves continue to be shot in considerable numbers on migration.

They are at the edge of their range here, and are nowhere abundant. Numbers fluctuate from year to year but an average density of 2–5 pairs per occupied tetrad is exceeded in only a very few favoured areas, including some river valleys. A Turtle Dove population of 250–600 pairs seems reasonable, with actual numbers probably at the lower end of this estimate in most years.   *GT*

CUCKOO
Cuculus canorus

Status: **Summer visitor**

Tetrads with evidence of breeding

Confirmed	83 (10%)
Probable	379 (44%)
Possible	321 (37%)
Total	**783 (90%)**

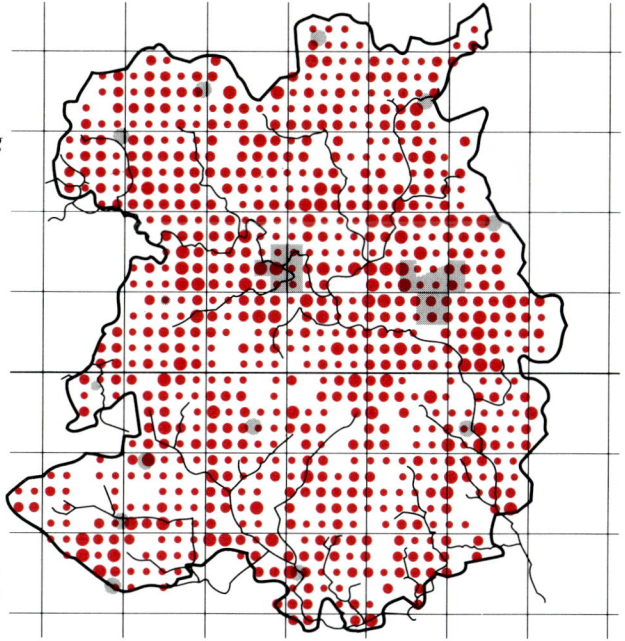

Hearing the first Cuckoo of spring is an eagerly awaited delight. They occasionally arrive in March, but are not widespread until the fourth week in April. Adults leave early, rarely being seen after July, although juveniles may stay throughout August, and occasionally into early September.

Habitat suitable for host species occurs in all tetrads. Shropshire Cuckoos are capable of parasitising about 19 species ranging in size from Wren to Blackbird, although the three main hosts are Dunnock, Meadow Pipit and Reed Warbler. The habitat of the last named is restricted, but Dunnocks are widespread in lowland woods, farmland and habitations, whilst Meadow Pipits are spread across open hills and moors.

Establishing that Cuckoos probably breed is not difficult, usually by observing a pair at a suitable site, or locating a permanent territory by hearing the characteristic far-carrying call. Confirming breeding is less easy, despite nestlings and fledglings being so noisy. Youngsters are often not seen until they are independent and away from their natal tetrad. The large number of possible breeding records will relate to males heard once in areas not revisited during the calling phase, or at the edge of their territory.

The expected correlation between Cuckoo and Meadow Pipit is not well demonstrated by the maps. In the latter's stronghold in the south-west, they occupied 34 tetrads in which the Cuckoo was unrecorded or only possible, implying that the Cuckoo was under-recorded in this remote and infrequently visited terrain. Hosts reported were Dunnocks (9), Reed Warblers (8), Tree Pipit, Meadow Pipit, Wren, Yellowhammer and Reed Bunting (*SBR*s 1985–90).

The national population trend is uncertain and information is conflicting. Hard evidence is difficult to find, although general opinion suggests a decrease (*Pop. Trends* 1990). Forrest (1899) reported them as "very abundant", and the *Handlist* (1964) as "widespread throughout the county . . . Reports indicate a decrease in recent years on low ground, but numbers appear to be maintained in hill country."

The Cuckoo's far-carrying call gives an impression of abundance, but they are thinly distributed and territories may cover several tetrads. Promiscuous breeding behaviour makes it better to refer to laying females rather than pairs, which, if the density of 5–10 per 10-km square applies (BTO *Atlas* 1976), gives a population of 175–350.   *JS*

BARN OWL
Tyto alba

Status: **Resident**

Tetrads with evidence of breeding

Confirmed	85 (10%)
Probable	58 (7%)
Possible	120 (14%)
Total	**263 (30%)**

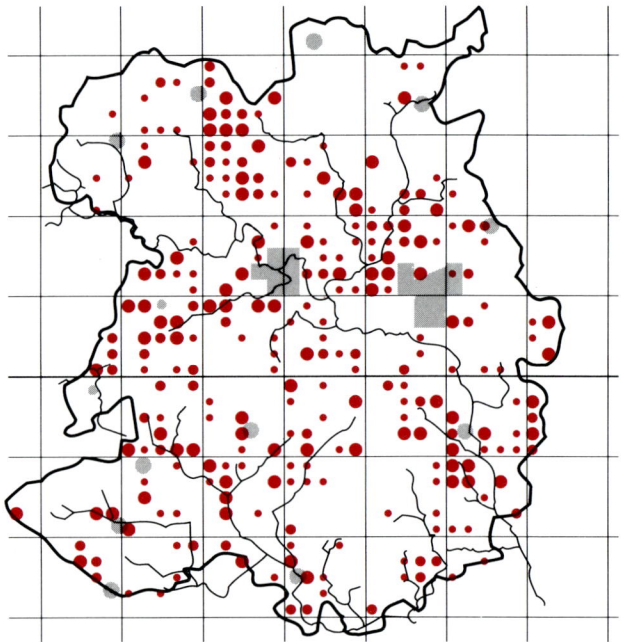

The ghostly form of the Barn Owl, floating buoyantly through the darkness, is more likely to be picked up by motorists' headlights than birdwatchers' binoculars, although it may be seen hunting in daylight, especially in winter or when feeding young.

Territories may be located by listening for the characteristic shrieks, but could have been easily overlooked, particularly in the more remote tetrads that were unlikely to be visited at dusk and night. Breeding is confirmed by watching for food-carrying adults at dusk or hearing the "snoring" of owlets in nests, which are usually in agricultural buildings, occasionally in derelict structures, and frequently in hollow trees, with elm, oak and ash most used. Many casual sightings have resulted in a high proportion of possible breeding records.

Forrest (1899) described it as "The most plentiful of the owls in Shropshire . . . gradually increasing in numbers". The *Handlist* (1964) stated it was well established, but "does not now merit H.E. Forrest's description" with "some evidence of a decrease in the north of the county in the past five years". Nationally, numbers have been dropping since the 1930s, but a sharper decline started in the 1940s when a series of severe winters, followed by organochlorine poisoning, reduced the population (*Pop. Trends* 1990). This loss has accelerated in recent years, although the issue is clouded by marked annual fluctuations in population related to winter weather and cycles in the abundance of voles. The main reasons for the demise of this once-common species are loss of prey-rich foraging areas, destruction of traditional breeding sites, urbanisation, use of toxic pesticides and disturbance (*Red Data* 1990).

In Shropshire the Barn Owl is still widely, but thinly, distributed, although it is absent from dense woodlands, built-up areas and a large part of the north-east, where it is much more scarce than the Little Owl, even though competition with the smaller species should have no significant effect. It has also fared badly due to road construction. Within two years of the Oswestry by-pass opening in 1986, 18 casualties were received by the RSPCA, of which only five were rehabilitated (B. Williams, pers. comm.). After adding corpses, the number killed and injured there was estimated at 40–50 in less than three years (*SBR* 1989). In 1980–90, 33 were reported killed on roads in other localities, and this is probably only the tip of the iceberg.

Although occasionally seen hunting in open woodland, it avoids competition with the dominant Tawny Owl by choosing suitable open habitats, now mostly reduced to riversides, road verges, traditional estates and parklands, and newly planted commercial woodland. Of the 143 tetrads occupied by Barn Owls, Tawny Owls were not recorded in 44 (30.8%). Daylight-hunting Kestrels share a similar diet of field vole, common shrew and wood mouse with the Barn Owl, but there is no apparent correlation between their distribution maps.

The Barn Owl is unlikely to regain former high numbers as long as current agricultural practices remain, although it could be encouraged by setting aside rough unimproved grassland and providing nestboxes in good foraging areas.

The increase in occupied 10-km squares since the BTO *Atlas* (1976) is almost certainly due to improved coverage. Shawyer (1987) described a drop from 287 pairs in 1932 to an estimated 95 pairs, but this seems low, even allowing for hunting owls being recorded in more than one tetrad, or pairs nesting in different tetrads during the six years of fieldwork. One pair per occupied tetrad would give a population now estimated at over 140 pairs.   *JS*

LITTLE OWL
Athene noctua

Status: **Resident**

Tetrads with evidence of breeding

Confirmed	214 (25%)
Probable	139 (16%)
Possible	232 (27%)
Total	**585 (67%)**

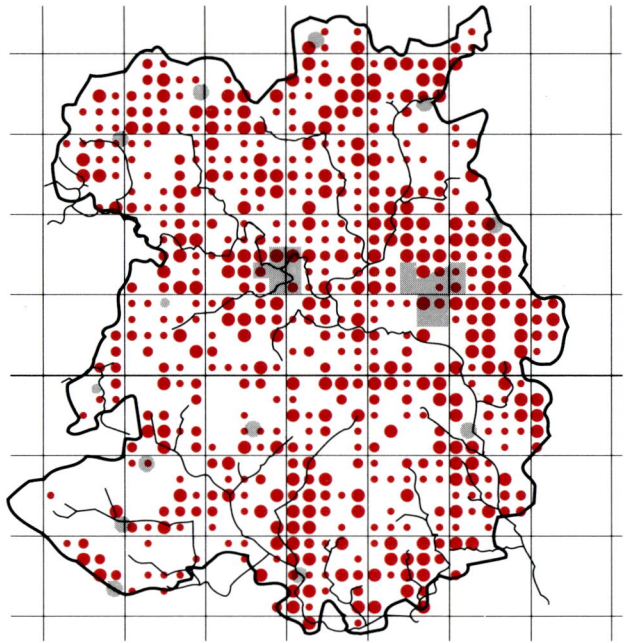

Sitting on a roadside post, compact and plump, with low forehead, flat crown and pale "eyebrows", the Little Owl appears to stare with disapproval at the world.

Introduced to England late last century, and first recorded in Shropshire in 1899, it was eventually confirmed breeding at Stanton Lacy in 1919. "By 1940 it was considered to be possibly the commonest owl", although numbers were thought to have decreased from the early 1950s (*Handlist* 1964).

Competition with the Tawny Owl is lessened by use of mainly open agricultural country and the avoidance of dense woodland. It also inhabits parks, large gardens and small habitations, but is absent from the large towns and uplands. Nests are usually in roomy tree holes, causing competition with two other predators, Barn Owl and Kestrel. It will use other sites, even nesting below ground in rabbit burrows.

The habit of perching near the nest in daylight makes visual location of territories easier than for other owls, and noisy territorial calls at dusk also betray sites. Pairs may remain faithful to a nest for many years. Numerous sightings of individuals in suitable habitat have resulted in a disproportionate number of possible records, many of which will relate to breeding pairs, and absence of evening and night visits to many tetrads suggests under-recording.

Although they are known to take large amounts of invertebrate prey (Mikkola 1983), wood mice and voles dominate the diet in years of abundance, probably accounting for a cyclical series of peaks and troughs in the national population since 1963 (*Pop. Trends* 1990). Numbers may also be affected by severe winters, although records submitted to the County Bird Recorder were lower for the latter years of fieldwork than for the earlier period which contained more adverse winter weather, in keeping with a national decline following a peak in 1984. Breeding is now confirmed in all the 10-km squares mainly in Shropshire, except SO37, whereas eleven of these 10-km squares were shown as unoccupied in the BTO

Atlas (1976). Though due in part to improved coverage, this also reflects an expansion in the distribution, and almost certainly the population, since then.

Shropshire contains much ideal habitat, and some tetrads hold several pairs. Territories are usually quite small (Mikkola 1983), and three territories were located in 0.25 sq. km in the Much Wenlock area. An average density of 1–3 pairs per recorded tetrad would give a population of 600–1800 pairs.   *JS*

TAWNY OWL
Strix aluco

Status: **Resident**

Tetrads with evidence of breeding

Confirmed	237 (27%)
Probable	197 (23%)
Possible	219 (25%)
Total	**653 (75%)**

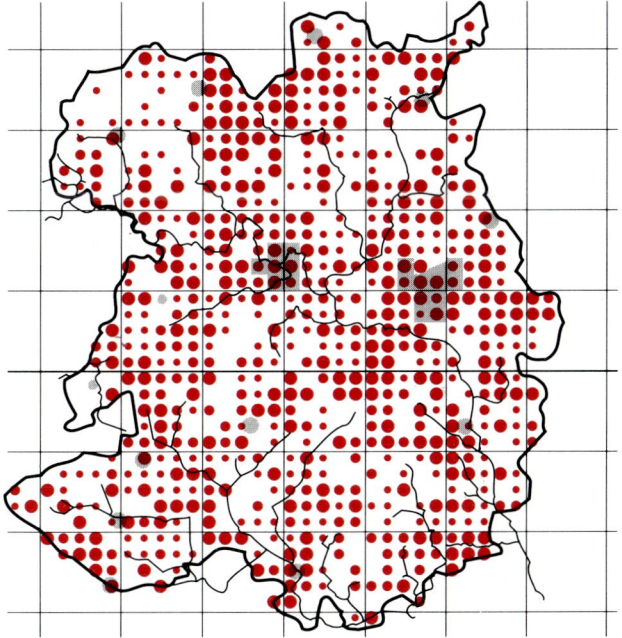

The melancholic call of the Tawny Owl is one of the best-known sounds, yet the bird is seldom seen. A rotund form perched on a roadside post or wire, silhouetted against a darkening sky or illuminated briefly by vehicle headlights, is the best sighting most people achieve. The choice of roadside habitat for hunting is often fatal, resulting in many casualties, and 17 deaths have been reported in the past 10 years (*SBRs*).

Tawny Owls are found in woodland, tree-scattered farmland, large gardens, churchyards, parks and built-up areas, where they nest mainly in tree holes and take readily to nestboxes. They will occasionally use large nests of other species, usually Carrion Crows. They are more common at low altitudes, but appear to avoid many small woods, and are often absent from those of less than 100 ha (*Pop. Trends* 1990). In larger woods territories are smaller and contiguous leading to much vocal competition, but less interaction on the much larger agricultural territories may have resulted in pairs being overlooked, and many of the possible records are likely to indicate breeding pairs, especially in the north-west. Poorly recorded in many areas with apparently suitable habitat, lack of evening and night visits must account for many blank tetrads. Owlets are frequently discovered in daytime, often away from cover, and this may explain the high proportion of confirmed records.

Reduced persecution from gamekeepers allowed the Tawny Owl to increase steadily from 1900 to 1930, with further local increases until 1950 (*Pop. Trends*

1990). Pairs remain within the territory throughout the year and pair for life (Mikkola 1983), are long-lived, can survive harsh winters and have a broad diet, which probably accounts for the general population stability since 1960, though numbers do fluctuate from year to year (*Pop. Trends* 1990).

The BTO *Atlas* (1976) suggested 50–100,000 breeding pairs in Britain, giving a range of roughly 25–50 pairs per occupied 10-km square. Applying this figure to Shropshire would give a population of around 900–1800 breeding pairs.

As Tawny Owls are very sedentary, and the population is regulated by territorial behaviour (*Pop. Trends* 1990), it seems likely to remain stable.   JS

LONG-EARED OWL
Asio otus

Status: **Resident**

Tetrads with evidence of breeding

Confirmed	6 (1%)
Probable	4 (0%)
Possible	10 (1%)
Total	**20 (2%)**

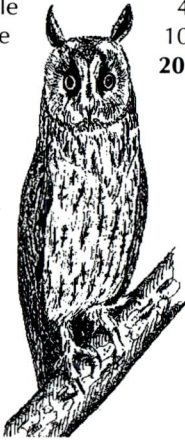

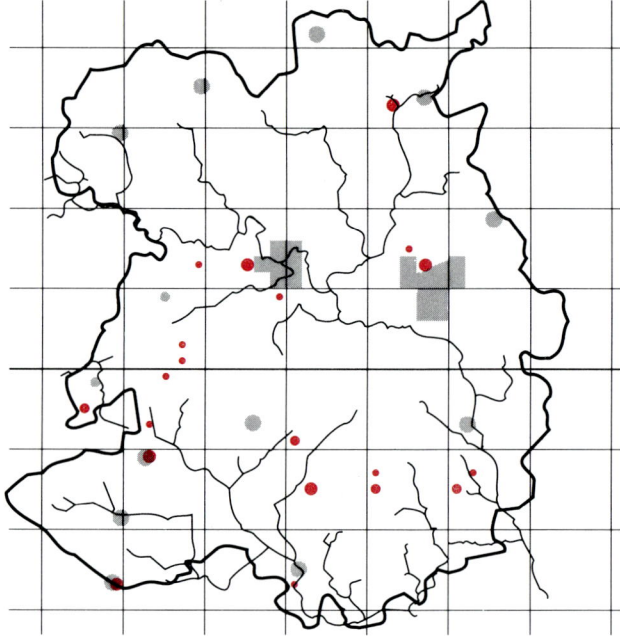

With its cryptic plumage and nocturnal rather than crepuscular habits, the Long-eared Owl is seldom seen. Roosting upright against a tree trunk, often in deep cover, it would still be overlooked even if it were not such a rarity.

The usual habitat is isolated plantations, shelterbelts, copses, thickets or overgrown hedges surrounded by open country, and they breed only on the edges of larger woods and forests (Mikkola 1983). Consequently they will not be so attracted to the large new forestry plantations of the south and south-west as commonly supposed. All tetrads with confirmed breeding contain only small woods of 4–8 ha except one near Knighton. Proving breeding is difficult, and requires night visits to locate territories by the quiet but penetrating "hoo-hoo-hoo-hoo" call, or the wing-clapping courtship flight, early in the year before fieldwork generally starts. Later, sites may be betrayed by the far-carrying discordant calls of the young.

Forrest (1899) described it as "Rather common, especially round The Wrekin. It prefers Fir woods." Since then it has declined drastically. The *Handlist* (1964) recorded it as apparently scarce, nesting regularly on The Long Mynd from 1955, and considered it to have increased in the Bucknell area, breeding usually "in old crows' nests in thorn scrub or fairly mature conifer plantations". It was not found in these areas during Atlas fieldwork. Over the past dozen years Long-eared Owls have been reported annually, most often out of the breeding season (*SBRs*).

The BTO *Atlas* (1976) showed Long-eared Owl present in up to 5 10-km squares, all in the south, compared with the present 15. This may be due largely to increased observer coverage, but there is an abundance of suitable habitat and nest sites, and a genuine increase in population and range cannot be discounted. The limiting factor is likely to be fierce competition from the Tawny Owl, which has similar ecological needs and is even known to kill its smaller rival (Mikkola 1983). The increase in Tawny Owls this century corresponds with the decline of the Long-eared Owls since Forrest's day.

Although under-recorded, assuming that all records refer to different territories would give a minimum population of 20 pairs.   JS

NIGHTJAR
Caprimulgus europaeus

Status: **Probable**

Tetrads with evidence of breeding

Confirmed	0 (	0%)
Probable	2 (	0%)
Possible	9 (	1%)
Total	**11 (**	**1%)**

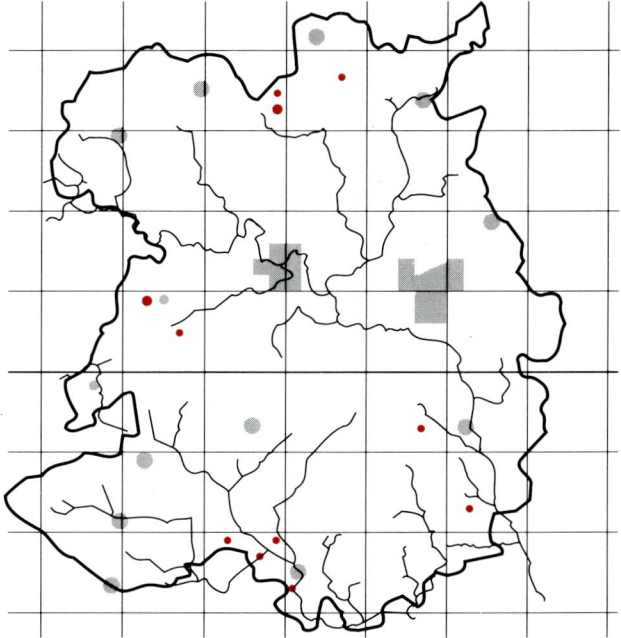

JS

This strange, crepuscular bird is now a rare summer visitor. Dry sandy heaths, recently felled woods and young conifer plantations are favoured but a wide variety of other habitats may be used. Though best located by listening for the male's territorial "churring" call around dusk, the Nightjar is easily overlooked and confirmation of breeding is very difficult to obtain.

A BTO Survey in 1957 found it was present in "fair numbers" on the northern mosses, at Haughmond Hill, and Willey Park, Broseley, and "numerous" in the young conifer plantations in the Teme valley (*SBR* 1957). Subsequently the *Handlist* (1964) recorded it as a "breeding visitor, small numbers present in all parts of the county". Since then the Nightjar has almost disappeared from Shropshire as part of a nationally recorded retreat from the north and west of its range in Britain, attributed to the destruction of heathland and other favoured lowland habitat, and climatic change, reducing the availabilty of insect food.

The BTO *Atlas* (1976) recorded them in seven 10-km squares, including confirmed breeding in two, but these were mostly single sites and not all occupied every year. Another BTO Survey in 1981 found none on the previously regular site at Whixall Moss, but churring males were noted on the Stiperstones (3), in Hawkstone Park (2) and along Wenlock Edge (2). Several other previously used sites were checked with negative results.

The decline has continued and the map overstates the position in any one year by combining the scatter of records throughout the Atlas period. There were no records in 1986 and 1989. The only 1988 record, of probable breeding, a pair and churring, came from woodland 4 km south of Whixall Moss but they were not reported there in later years. The 1990 records came from the extreme south apart from the second probable breeding record, a churring bird in scrubland at Vennington on the slopes of Long Mountain. The only heathland record was from Prees Heath in 1985, whilst other sites included young conifer plantations and mature deciduous woodland. None of the sites occupied in 1981 produced records for this Atlas, nor did The Wrekin, where two chicks found in 1983 provide the most recent record of confirmed breeding (*SBR*).

The population now fluctuates from nil to not more than five or six pairs annually.

CEW

SWIFT
Apus apus

Status: **Summer visitor**

Tetrads with evidence of breeding

Confirmed	290 (33%)
Probable	150 (17%)
Possible	265 (30%)
Total	**705 (81%)**

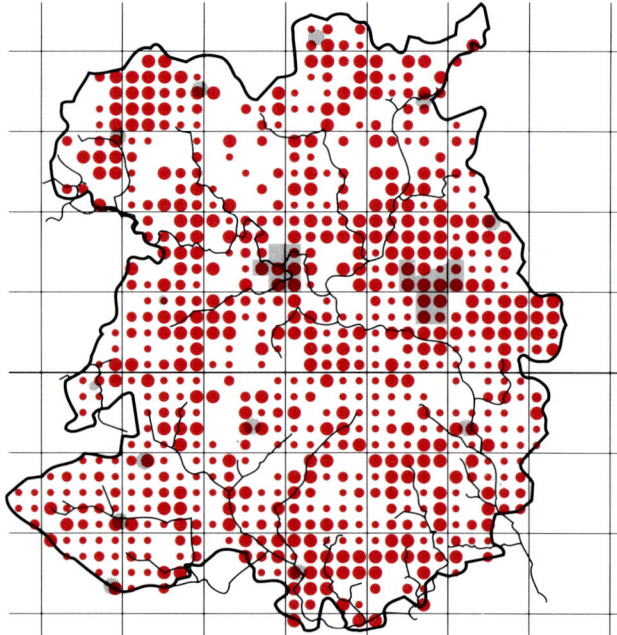

Parties of screaming Swifts, dashing follow-my-leader through towns and villages, are a feature of the summer months. The earliest arrive in late April, and most have left by mid-August, making their 16-week stay the shortest of all summer migrants'.

Swifts nest under eaves and in openings of old buildings, but in Shropshire they are not known to nest on cliffs or quarries as in some areas of Britain, or use tree holes as in some Continental countries. Old-established buildings are more likely to provide breeding sites than modern estates and office blocks. Site-fidelity is very strong in adults, but weak in the young (*BWP*), and if new sites become available the latter could utilise them. They will take readily to well-positioned nestboxes and may occasionally choose very low sites. In 1988 a pair nested in High Street, Much Wenlock, with the entrance under eaves only three metres above the pavement, and the parents dashed away at head-height when leaving.

Swifts travel long distances and may be seen anywhere, accounting for the high number of possible breeding records. Feeding parties are usually small, but flocks of several hundred may be seen, especially above The Long Mynd and other hills, and along Wenlock Edge. These are probably immatures as breeding is not usually successful until the fourth year (*BWP*). Some observers, noting Swifts in tetrads where there are no obvious nesting buildings, have neglected to record the species present, accounting for blank tetrads.

They are semi-colonial, and several nests may be served by one entrance. Breeding is usually confirmed by observing adults entering nest holes, and occasionally through finding youngsters stranded on the ground, having left the nest prematurely. As adults return there so infrequently many sites will have been missed, but some non-breeders entering holes may have been recorded as confirmed. No population change has been detected this century, but redevelopment may have reduced the number of nesting sites. The BTO *Atlas* (1976) estimated about 40 pairs per occupied 10-km square, and applying this figure to Shropshire would give a population of around 1400 breeding pairs. There are also large numbers of non-breeding immatures.

JS

KINGFISHER
Alcedo atthis

Status: **Resident**

Tetrads with evidence of breeding

Confirmed	81 (9%)
Probable	56 (6%)
Possible	117 (13%)
Total	**254 (29%)**

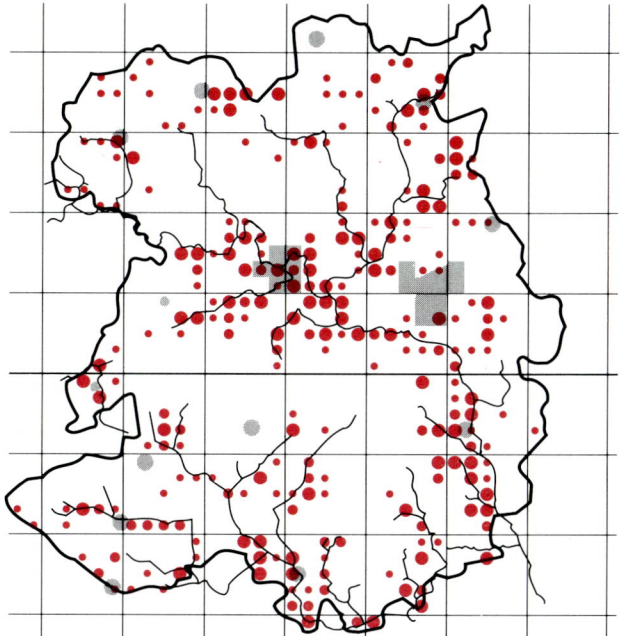

A high-pitched whistle or a flash of electric blue is often the first indication of this jewel of Shropshire's rivers and streams.

The slower-flowing rivers, including the Severn, Tern, Teme and Clun, provide breeding strongholds, although Kingfishers also breed on many tributary streams and brooks. Exposed vertical banks are essential for nest tunnels. Sometimes they are seen by lakes, reservoirs and canals during the breeding season, but nesting occurs infrequently in this habitat.

Once they have been found on territory in spring, proving breeding should not be difficult. A pair may raise three broods in a season, giving extended opportunities to note parents carrying food, or see recently fledged young. However, they are wary, and some waters are not easily accessible. This may account for the large percentage of possible breeding records not being upgraded, and other pairs may have been missed.

Breeding populations fluctuate according to the harshness of preceding winters. The *Handlist* (1964) stated that the Kingfisher was badly hit by the severe 1962–63 winter when it was exterminated in some areas, and countrywide it was more badly affected than any other British bird (*Pop. Trends* 1990). Several successful breeding seasons followed, which resulted in recolonisation of affected waters over the next few years. Mild winters over the latter period of Atlas fieldwork have restored numbers to a higher level, following a drop due to the harsh weather in early 1985 and a month of continuous freezing throughout February 1986, reflecting the national situation (*Pop. Trends* 1990). Compared with the BTO *Atlas* (1976), Kingfishers have expanded their range slightly, the most significant advance being along the River Corve.

Water Bird Surveys in prime habitat on the River Severn near Shrewsbury have found nests to be around 2–3 km apart (C.E. Wright, pers. comm.). Meanders or tributaries mean some tetrads have several kilometres of waterway and support 2–3 pairs, while other territories include part of two tetrads. An average of 1–2.5 pairs per occupied tetrad gives a population estimate of 140–350 breeding pairs.

In recent years disturbance by anglers from the start of the coarse fishing season has resulted in the desertion of some nests (*SBR* 1987), and absence of breeding pairs on some streams may be due to pollution. *JS*

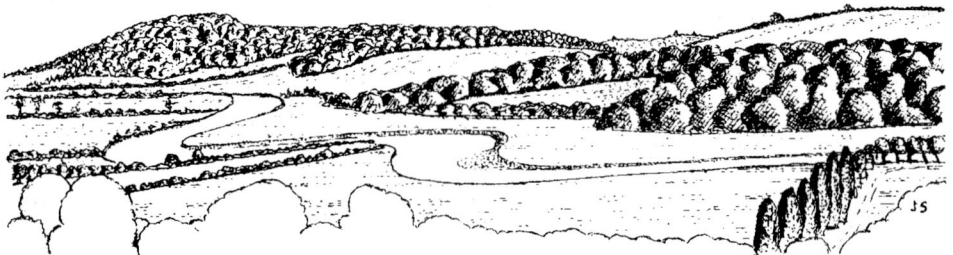

GREEN WOODPECKER
Picus viridis

Status: **Resident**

Tetrads with evidence of breeding

Confirmed	132 (15%)
Probable	154 (18%)
Possible	204 (23%)
Total	**490 (56%)**

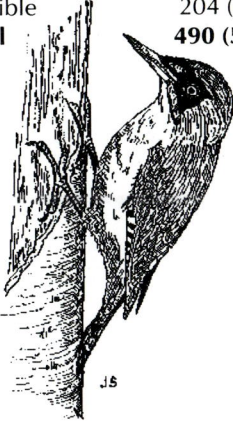

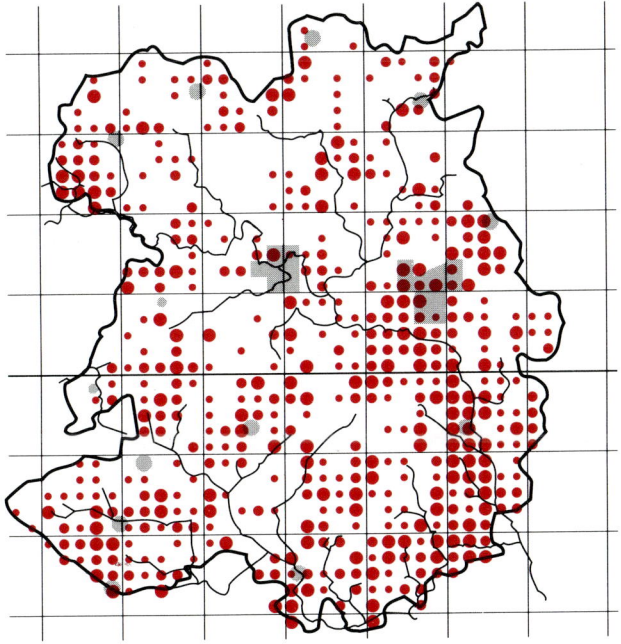

The loud, far-carrying "yaffle" call is often the first clue to the presence of this, the largest and most colourful woodpecker.

The Green Woodpecker is typical of wooded areas with open ground, and parkland and farmland with mature trees. More open areas such as heathland, grazed commons and golf courses are also favoured, so long as there are suitable trees nearby in which to nest. These open areas provide habitats for the turf-dwelling invertebrates on which it mainly feeds, and individuals have been seen to fly repeatedly over 600m to collect food from such productive areas. Feeding on wood ants in mature deciduous and coniferous woodland is common, but nesting would be unlikely in young even-aged stands of the latter. Dead or dying trees in which to excavate a nest are needed, although existing holes are readily used.

The Green Woodpecker is widespread but irregularly distributed, with breeding strongholds in wooded areas of the lower Severn valley, the Oswestry uplands, the Clun Forest and the Clee hills. It also occurs in the upland areas of the Stiperstones and The Long Mynd. Elsewhere it is more thinly distributed, particularly in north-central areas, due to lack of suitable nesting and feeding sites.

In early spring the characteristic territorial call makes presence easy to detect, but it becomes much less vocal later in the season, and could have been missed then. The low percentage of confirmed breeding records is due to the occupation of large territories, which makes nests hard to find, and fledglings being with their parents for only a short period. Though present in many areas at very low densities, any under-recording may have been compensated for by recording some pairs in two tetrads.

Of the three woodpecker species, the Green Woodpecker suffers most from severe winters (*Pop. Trends* 1990), and the *Handlist* (1964) stated that "Its numbers were very much reduced by the severe winter of 1962–63". It is also susceptible to changes in land use but was probably the least affected by Dutch elm disease (Osborne 1982). The population is likely to be higher now than that indicated by the BTO *Atlas* (1976), as a result of a full recovery following the winter of 1962–63 and increased conifer afforestation providing wood ant habitats. This growth in population may have resulted in a range extension, accounting for breeding now in five of the six 10-km squares unoccupied in the BTO *Atlas* (1976).

On a reasonable assumption of 1–2 pairs in each occupied tetrad and some of the possible breeding records representing overlooked pairs in less-favoured areas, the population is estimated at around 500–1000 pairs.

JM

GREAT SPOTTED WOODPECKER
Dendrocopos major

Status: **Resident**

Tetrads with evidence of breeding

Confirmed	300 (34%)
Probable	207 (24%)
Possible	211 (24%)
Total	**718 (83%)**

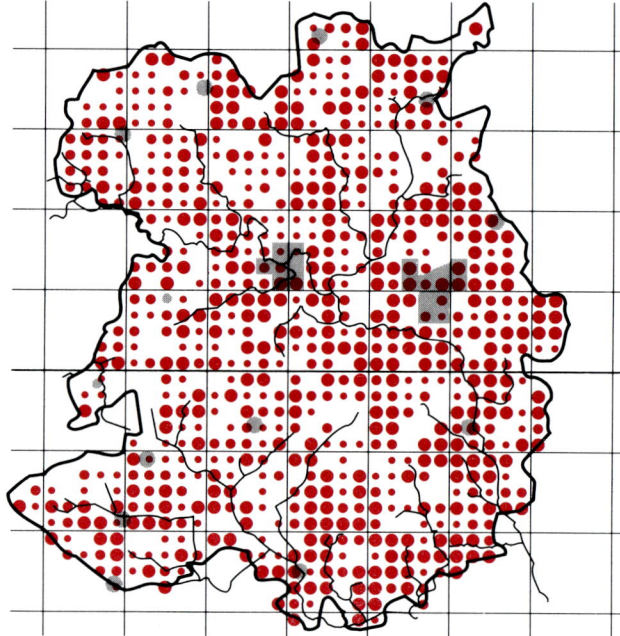

The Great Spotted Woodpecker is the most widely distributed of the three resident woodpeckers, but the one most demanding of woodland as a habitat.

Territorial "drumming" in early spring and the characteristic "tchick" call make initial location relatively easy, and nest-hole excavations with wood chips lying at the base of a tree provide further evidence of occupation. Well-grown young in the nest are very noisy, adults carrying food can be followed back after foraging within a few hundred metres of the nest site, and red-capped juveniles can be separated from adults, so breeding is confirmed more easily than for the other woodpeckers.

The optimum habitat is mature mixed woodland with dead and dying trees, especially birch, which provides suitable nest sites and food. It is also found in old orchards, parkland and wooded golf courses, and small woodlands are used if there are tree corridors connecting them to other woods, or if they are within commuting distance, which may be up to 800m. Young, even-aged conifer plantations and open agricultural areas with few trees are usually avoided. Otherwise it is widespread, being found at all altitudes and extending into well-wooded farmland in the north and east. Many of the tetrads from which it appears to be absent lack woodland of suitable type and size.

The BTO *Atlas* (1976) recorded the Great Spotted Woodpecker as absent from SO59 and only as present in 5 other 10-km squares, 4 in the south. Breeding has now been confirmed in four or more tetrads in each of these 10-km squares, suggesting that there has been a range expansion together with an increase in the population, probably in response to the outbreak of Dutch elm disease (Osborne 1982). However, numbers probably peaked during the early years of Atlas fieldwork and any decline is more likely to be attributable to the subsidence of Dutch elm disease rather than the effect of hard winters (*Pop. Trends* 1990).

Densities can be in the order of 10 pairs per sq. km in optimum areas but only 1 pair per tetrad in marginal habitats. Mainly a sedentary species, possible and probable records are likely to refer to breeding pairs, and, assuming an average of 2–4 pairs per tetrad, the population is estimated at around 1500–3000 pairs.   JM

LESSER SPOTTED WOODPECKER

Dendrocopos minor

Status: **Resident**

Tetrads with evidence of breeding

Confirmed	51 (6%)
Probable	57 (7%)
Possible	107 (12%)
Total	**215 (25%)**

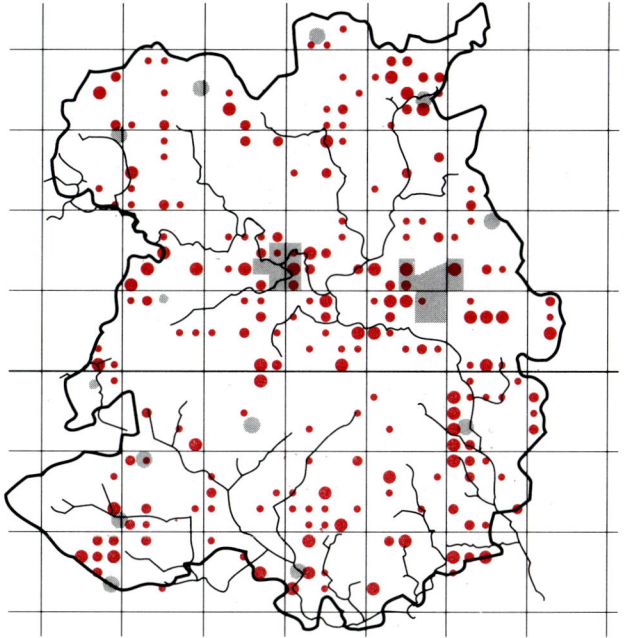

Small size, elusive behaviour and scattered distribution all contribute to make this the scarcest and least well-known woodpecker.

The Lesser Spotted Woodpecker overlaps the Great Spotted Woodpecker for much of its range, especially in mature deciduous woodland, but avoids conifer plantations. Otherwise, more diverse breeding habitats are occupied, including small woods, overgrown hedges with trees, tree corridors along streams and rivers, and other damp areas, especially with alders.

Initial location is not easy. Territorial "drumming" and the typical call — "pee-pee-pee . . ." of seven or eight notes — can be heard in early spring, but both occur only for a very short time, especially where densities are low and pairs are widely separated. It forages in the tree-tops for much of the time and is usually overlooked when the canopy is fully developed. Well-grown nestlings are very vocal for a short period, and chance observation of adults carrying food may also identify nest sites, but these are rarely discovered.

Distribution is scattered, but localised, showing the difficulty of obtaining confirmed breeding. Though found above 300m, the Lesser Spotted Woodpecker predominantly occupies lowland areas, particularly river valleys such as those of the Severn, Teme, Clun, Onny and their tributaries, and the headwaters of the Tern and Roden in the north.

The *Handlist* (1964) described this woodpecker as a "local but thinly distributed resident in lowland areas". The BTO *Atlas* (1976) showed the main breeding areas were in the north and north-east with confirmed breeding in only 9 10-km squares. It was not found in 13 10-km squares, generally in the south and west. The present Atlas shows confirmed breeding in 28 10-km squares. This apparent extension of range may be due to greater fieldwork effort, but is much more likely to be a genuine expansion following increased availability of nest sites and food in the wake of Dutch elm disease (Osborne 1982). Now this epidemic has died out numbers may be falling again (*Pop. Trends* 1990).

A high proportion of possible and probable records are likely to refer to breeding pairs, and many others will have been overlooked; thus, taking a reasonable average of 1–2 pairs per tetrad, the population is estimated at around 250–500 pairs.   JM

SKYLARK
Alauda arvensis

Status: **Resident**

Tetrads with evidence of breeding

Confirmed	224 (26%)
Probable	377 (43%)
Possible	163 (19%)
Total	**764 (88%)**

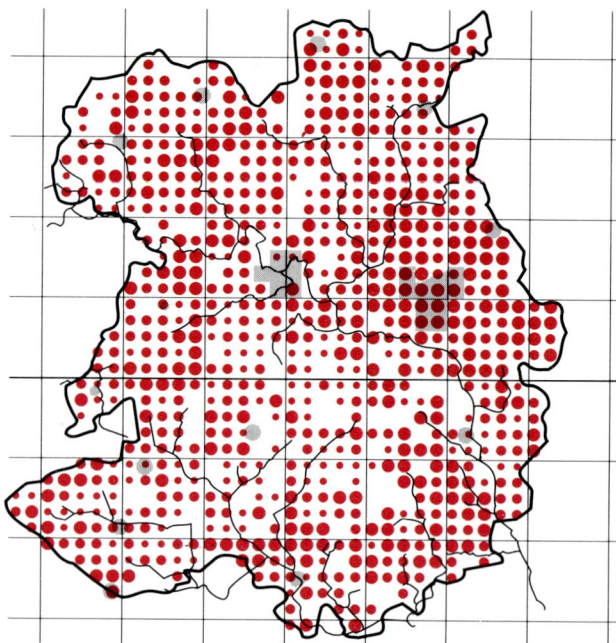

The resident and largely sedentary Skylark is gregarious in winter, when flocks are supplemented by Continental visitors. In summer it can be found in any open country, ranging from moorland and upland grass to lowland pasture and cereals, singing continuously in its familiar display flight for long periods, often so high as to be almost out of sight. Finding the nest, however, is a matter of luck or extreme vigilance, though adults carrying food may provide evidence of confirmed breeding. Hence very nearly half the records are probable breeding, while possible breeding records almost certainly relate to nesting pairs in tetrads with a low population which were visited too infrequently to prove a territory.

Due to changing farming methods which affect both winter and breeding conditions, the Skylark population is now declining rapidly in Britain and fell by almost 50% on CBC farmland sites in only eight years from 1980 (*Pop. Trends* 1990). The same changes — a shift to autumn-sown cereal, with less winter stubble and spring plough; poorer crop diversity; increasing use of herbicides, decreasing weed abundance; and increasing urbanisation — are all taking place in Shropshire too. This large-scale creation of unsuitable habitat is the most likely explanation for the gaps in distribution. Local people familiar with their area believe that Skylarks are now less common, and the description in the *Handlist* (1964) — "common throughout the county" — certainly no longer applies.

If the population decline is as severe here as nationally, the map probably gives a misleading impression. The broad spread of records may not mean high numbers, and some tetrads occupied at the start of the Atlas may no longer hold Skylarks. The national population was estimated at 2 million pairs in 1982, suggesting an average then of just over 700 pairs per 10-km square. As western England has a relatively low density and the population has fallen substantially since 1982 (*Pop Trends* 1990), it may now be as low as 400 pairs per 10-km square, a total of around 14,000 pairs, or 25 pairs per occupied tetrad.   *DS*

SAND MARTIN
Riparia riparia

Status: **Summer visitor**

Tetrads with evidence of breeding

Confirmed	64	(7%)
Probable	18	(2%)
Possible	35	(4%)
Total	**117**	**(13%)**

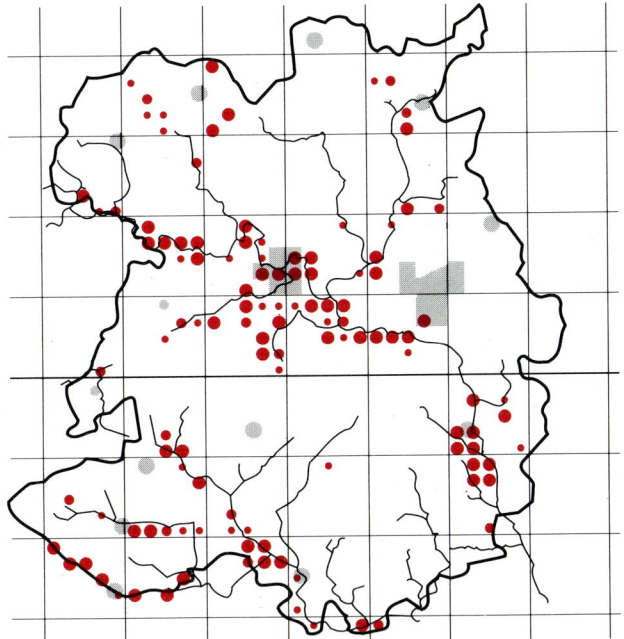

The British population of Sand Martins reached its lowest level for at least twenty years in 1984, just prior to the start of the Atlas. Monitoring by the BTO and others witnessed a dramatic crash in the number of spring arrivals in 1969, compared to 1968, followed by a further slower subsequent decline ending in 1984 when only 10% of the 1968 population remained. Drought in the wintering quarters in the Sahel region of central and West Africa is thought to have been the prime cause, though a series of cool springs in Britain also reduced breeding success. A recovery has taken place in the late 1980s, but numbers returning each spring vary considerably from year to year mainly in line with conditions in the Sahel, and remain much lower than in the 1960s (*Pop. Trends* 1990).

Shropshire numbers have paralleled national trends with an increase in reports since 1985 (*SBRs*). However, confirmation of breeding has only been obtained from 9 of the 16 10-km squares wholly in Shropshire in which breeding was proved in 1968–72 (BTO *Atlas* 1976).

Nest burrows are excavated in vertical sand banks, with some of the largest colonies in man-made earthworkings or quarries such as Venus Pool, Hilton, Condover and Wood Lane. Naturally eroding riverbanks are the original habitat, with certain stretches of the Severn especially suitable, notably between Shrewsbury and Buildwas. Other riverbank colonies are known on the Tern, Teme, East Onny and West Onny. When the population was high, holes were even dug in sandstone cliffs exposed in the construction of the Shropshire Union Canal at Tyrley, Market Drayton, but these have been abandoned since the early 1970s.

Such a specialised habitat makes location of breeding colonies easy, though small isolated ones may have been overlooked, especially in tetrads not visited every year. Where they are found, confirmation is easy and the young can even be heard within their burrows, but colonies in both man-made and natural sites are transient and may disappear from one year to the next, often relocating at an alternative site nearby. For example, the former major colony at Venus Pool, which had over 300 nest holes in 1983 when the general population was very low, was deserted in 1985 and 1989 but occupied in much lower numbers in the intervening years (*SBRs*).

Groups of over 100 holes are now scarce and assuming one colony per occupied tetrad and an average of about 50 pairs per colony, a population of around 4000 pairs is estimated for the end of the Atlas period.

Droughts in the Sahel appear to be the major limiting factor on the population, though obviously good breeding success is vital if there is to be any chance of a sustained recovery. However, as fewer licences for the extraction of sand and gravel are now being granted, and older workings are in-filled or otherwise reclaimed, these man-made habitats will be reduced. Retention of the natural character of rivers and especially riverbanks is therefore vital to ensure the welfare of the Sand Martin, as anti-erosion schemes would reduce available natural habitats.   GT

SWALLOW
Hirundo rustica

Status: **Summer visitor**

Tetrads with evidence of breeding

Confirmed	795 (91%)
Probable	46 (5%)
Possible	19 (2%)
Total	**860 (99%)**

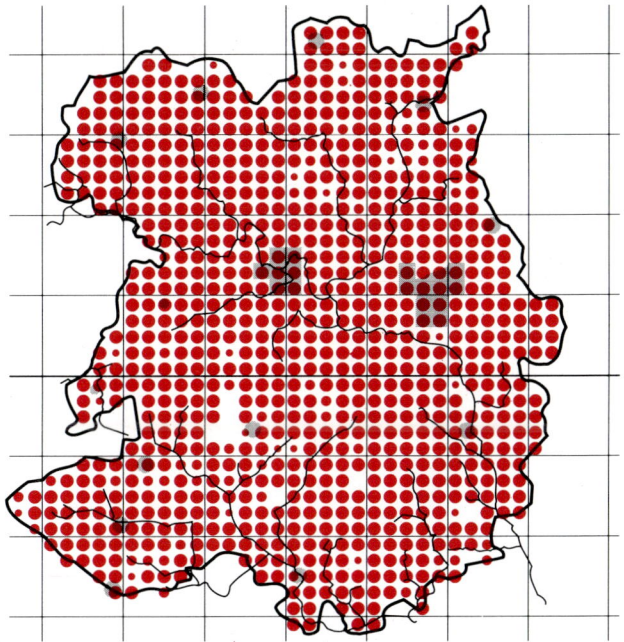

Perhaps the most familiar bird of summer, the Swallow is closely associated with human habitation, especially farms, villages and other rural communities. It arrives in strength in April, and breeding starts almost immediately. Barns, milking sheds, other buildings and even the porches of houses and churches are used, with the relatively bulky nest usually being placed on a ledge or beam, or in another equally obvious position. The noisy alarm calls of adults, the typical selection by newly fledged chicks of prominent perches such as overhead wires, and the usual raising of two broods, sometimes three, all help to locate and confirm breeding. The nesting season can extend as late as September, though only one such occurrence has been recorded recently, a nest found at Worthen on 3 September 1979, containing three eggs (*SBR*).

Most records are therefore of proven breeding. Swallows are scarcer in upland areas but nevertheless are usually present where quiet buildings are available, and only the tetrads of The Long Mynd lack suitable nest sites.

Swallows have decreased in abundance considerably in the last decade. Various causes have been suggested — primarily drought in the winter quarters in South Africa, and, to a lesser extent, changes in farming practice here, which have resulted in fewer buildings with suitable nest sites and a reduction in breeding areas for flies and other insect food. The Sahel drought may also be a factor as it has effectively widened the desert, making the trans-Saharan migration more hazardous, so Swallows arrive here in poorer condition for breeding (*Pop. Trends* 1990).

The BTO *Atlas* (1976) estimated an equivalent of 5–10 pairs per tetrad nationally, but numbers have decreased by at least 30% since then. Assuming that these estimates apply in Shropshire, a population of 3000–6000 pairs is suggested.   *GT*

HOUSE MARTIN
Delichon urbica

Status: **Summer visitor**

Tetrads with evidence of breeding

Confirmed	758 (87%)
Probable	27 (3%)
Possible	47 (5%)
Total	**832 (96%)**

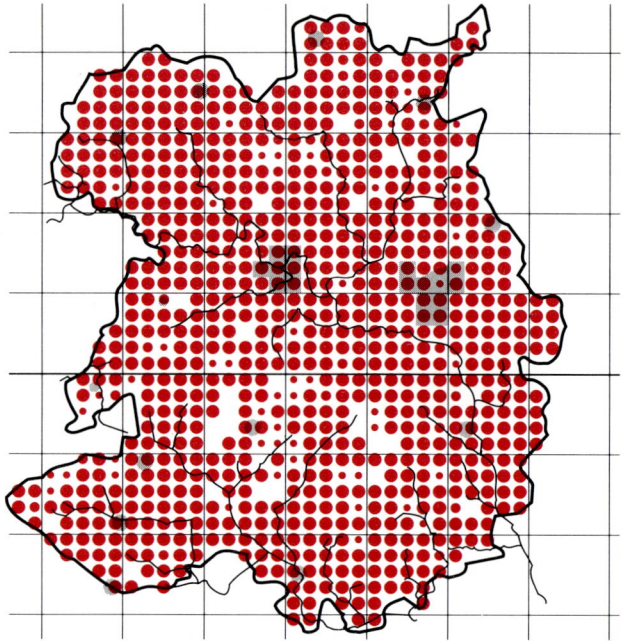

Few species present so little difficulty for the Atlas worker: House Martins are conspicuous and easy to identify, and are confiding when visiting their very obvious nests during a protracted breeding season.

There are few tetrads where breeding has not been proved, the most notable absences being in tetrads covering The Long Mynd, which are devoid of nest sites. Other lean areas include the Stretton hills, the Wyre Forest, intensive arable farmland, and sparsely populated parts of the south-west.

House Martins were originally cliff nesters, and remain so in several areas, but this habit has never been recorded in Shropshire. There are few suitable cliff sites and it was possibly not until the construction of buildings and bridges that House Martins became established.

Colonies generally appear to be larger and more numerous in the suburbs and villages of the north and east, especially in mixed farming areas, than in the south and west where buildings are more scattered; altitude and humidity may play a part too. *Pop. Trends* (1990) cites two nests per sq. km as a possible average, which suggests a total population of some 7000 pairs.

The best known colony is at Atcham old bridge. Forrest (1908) referred to the nests being in "an almost continuous row . . . from end to end". Counts by Dr Bruce Campbell showed 168 occupied nests in 1957 rising to 325 in 1962 (*SBR*s). There were few counts made in the 1970s and the highest was only 71; the highest since then was 143 occupied nests in 1984. Such fluctuations in the size of an individual colony are not unusual and are an unreliable guide to population trends over a wider area (*Pop. Trends* 1990). Campbell speculated that the Atcham old bridge colony was the largest known in Great Britain but subsequently a colony in excess of 500 pairs has been recorded in Oxfordshire (*BWP*).

A bonus for the Atlas worker is that House Martin nests also frequently provide breeding records of House Sparrow. Other more unexpected occupants during the Atlas years were Tree Sparrows at Beckbury and Stanmore, Wrens at Hopesay and Atcham bridge, where a nest was also used by Pied Wagtails, and Spotted Flycatchers at Hopton Cangeford (SBRs).   TW

TREE PIPIT
Anthus trivialis

Status: **Summer visitor**

Tetrads with evidence of breeding

Confirmed	76 (9%)
Probable	74 (9%)
Possible	73 (8%)
Total	**223 (26%)**

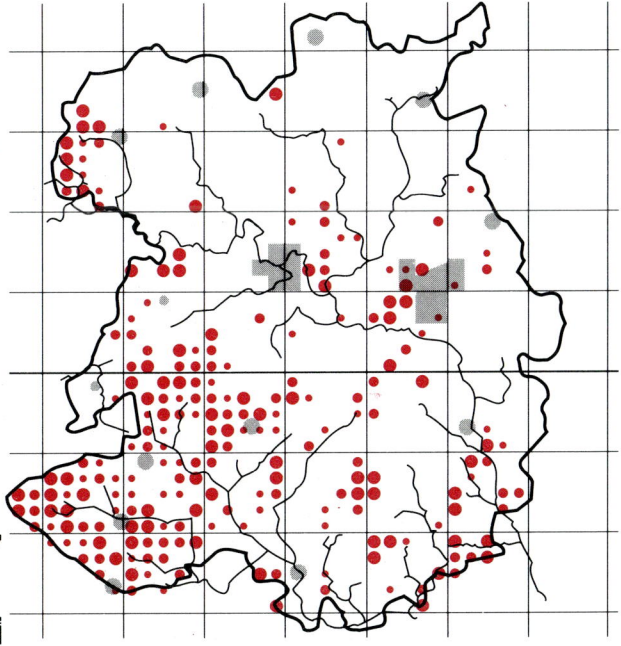

Tree Pipits winter in Africa south of the Sahel, returning from late March and becoming widespread by the end of April. Often found in the same tetrad as their close relative the Meadow Pipit, they occupy separate habitats. Tree Pipits nest in open woodland or rough areas with scattered trees or bushes from which they perform their distinctive song flights. These displays enable territories to be located easily, and breeding can be confirmed when young are being fed, as the adults call agitatedly from a prominent position and will not return to the nest while an intruder is present.

Traditional farming methods combined with the upland topography have maintained a mosaic of copses, wooded valleys and rough pasture with scattered trees on hillsides which is ideal for Tree Pipits. The populations of Clun Forest, Stiperstones, The Long Mynd and Stretton hills merge to form the large concentration in the south-west. Further east they are found on Wenlock Edge, Catherton Common and Brown Clee. To the north, The Wrekin, Haughmond Hill and Oswestry uplands are the main centres, with lower numbers on the smaller hills at Lilleshall, Edgmond, Nesscliffe, Middletown and Loton. In the south-east they occur in the Wyre Forest and some of the smaller woods nearby. Whixall Moss, represented by a single tetrad in the north, provides a different habitat where birch scrub is encroaching onto lowland bog.

Tree Pipits respond quickly to changes in the environment. They will nest in new plantations or clear-felled woodland, provided that suitable song posts are available, and leave again when conditions become unfavourable. Due to the transient nature of some breeding sites the map may slightly exaggerate the position in any one year.

Atlas fieldwork suggests an average of 6–12 pairs for each occupied tetrad, giving a population of around 900–1800 pairs. A CBC plot in the north-west only recorded 3 territories in 1989 and 2 in 1990 while it had averaged 8 during the previous four years (T.W. Edwards, pers. comm.). Hopefully this is merely a fluctuation not a trend.

APD

MEADOW PIPIT

Anthus pratensis

Status: **Resident**

Tetrads with evidence of breeding

Confirmed	88 (10%)
Probable	68 (8%)
Possible	87 (10%)
Total	**243 (28%)**

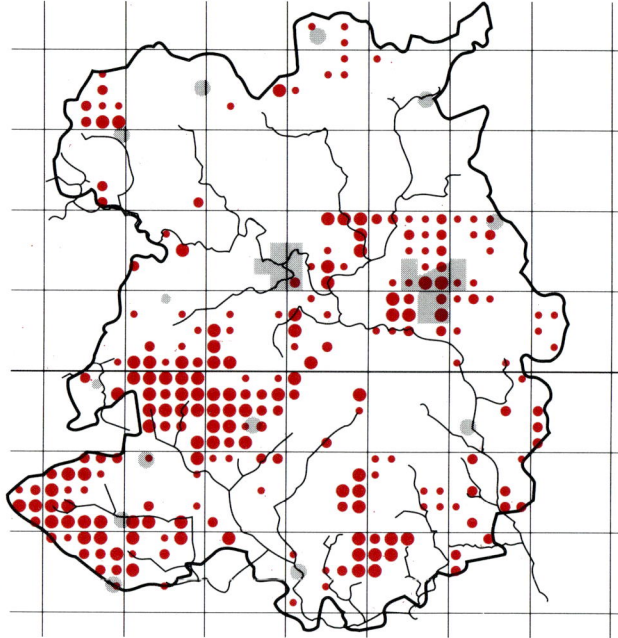

Most of our Meadow Pipits winter in southern Europe, returning from late March into April. Their song flight is similar to the Tree Pipit's but generally starts from and returns to the ground. Meadow Pipits occur chiefly in the uplands where heather moorland and rough pasture have been preserved. The open nature of the habitat enables adults to be seen returning to their nests with food from quite a distance and this, repeated for two or three broods, makes confirmation of breeding easy.

The distribution of the Meadow Pipit is quite similar to that of the Tree Pipit. Where scrub begins to invade an open site both may be present, as on Catherton Common and Whixall Moss, but the Meadow Pipit is absent from wooded areas in the Oswestry uplands, Wenlock Edge and the valleys which separate the southern hills. Additionally Meadow Pipits occur in small numbers in a wide arc around Telford, where habitats vary from rough grass on higher ground to the east, damp fields on peat soils to the north, and derelict areas which may be lost to construction work, such as "a breeding site at The Rock fast disappearing under houses" (*SBR* 1985).

An estimated 170 pairs were found on Brown Clee (*SBR* 1990). Away from their stronghold in the south-west, late wintering or migrant birds, or those driven from the hills by harsh spring weather, may be responsible for some records, but low densities in a marginal and limited habitat will also have made confirmation of breeding pairs more difficult.

Densities of 25–50 pairs/sq. km are common on moorland and upland sheepwalk (*Pop. Trends* 1990). Assuming a quarter of each occupied tetrad in the main area is suitable habitat there are probably 2500–5000 pairs. Nationally numbers declined in the early 1980s, reaching a low in 1986 before a slight improvement began (*Pop. Trends* 1990). This was reflected on a CBC plot near Oswestry where after recording 4 territories in 1980 only 1 was found each year in 1983–5, since when 4 have been found regularly (T.W. Edwards, pers. comm.).   *APD*

YELLOW WAGTAIL
Motacilla flava

Status: **Summer visitor**

Tetrads with evidence of breeding

Confirmed	146 (17%)
Probable	88 (10%)
Possible	82 (9%)
Total	**316 (36%)**

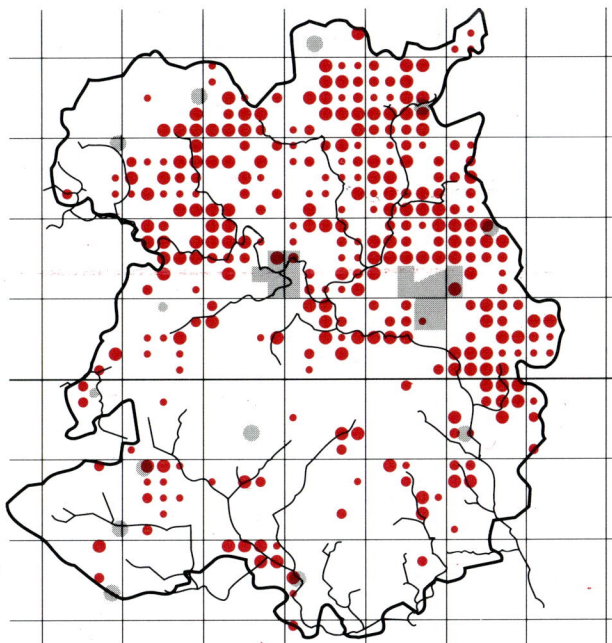

The Yellow Wagtail is most frequently recorded in spring and autumn as a passage migrant, usually from along the Severn valley. It breeds only in small numbers.

Atlas fieldwork has identified two distinct breeding habitats. Some nests are in damp pasture on the floodplains of the main rivers. Most are in fields of well-irrigated crops, particularly potatoes, though cabbage, sugar beet, peas, oil seed rape and cereals are also used.

The *Handlist* (1964) described the Yellow Wagtail as "fairly common in the river floodplains and marshy pasture throughout the county, but absent from hill country"; no reference was made to presence on cultivated land. Though the current map may only be the result of a systematic survey, nesting in crops is probably a recent adaptation after loss of much of the original habitat through agricultural drainage since the *Handlist* was published. Even if the habitat has changed, the range does not appear to have altered significantly since the BTO *Atlas* (1976).

The Yellow Wagtail map is one of the most fascinating in the Atlas. It shows excellent correlation with the combined areas of the two distinct breeding habitats — the valleys along the main rivers (Map 2), which is largely the land below 61m (Map 5); and the cultivated land (Map 7).

Local farming practice usually rotates cereals and root crops, so the Corn Bunting distribution map can also be taken as a good, more localised, indicator of cultivation. The two species use different habitats in the same field system, and the vast majority of tetrads occupied by Yellow Wagtails are either in a river valley or hold Corn Buntings (see Map 11).

Once Yellow Wagtails are located, confirming breeding is quite easy. When young are in the nest, or recently fledged, both adults perch prominently on lookout before returning to them with beaks full of food. Apparently suitable habitat is widespread, but many fields are not used. Those that are chosen often hold small loose colonies of several breeding pairs.

Some of the possible and probable breeding records will relate to passage migrants, particularly outside the main breeding range, though isolated nesting pairs are found. Numbers are also known to fluctuate considerably from year to year. Atlas fieldwork suggests occupied tetrads may contain an average of 5–10 pairs, giving a population estimate of around 1150–2300 pairs. *LS*

GREY WAGTAIL
Motacilla cinerea

Status: **Resident**

Tetrads with evidence of breeding

Confirmed	150 (17%)
Probable	74 (9%)
Possible	106 (12%)
Total	**330 (38%)**

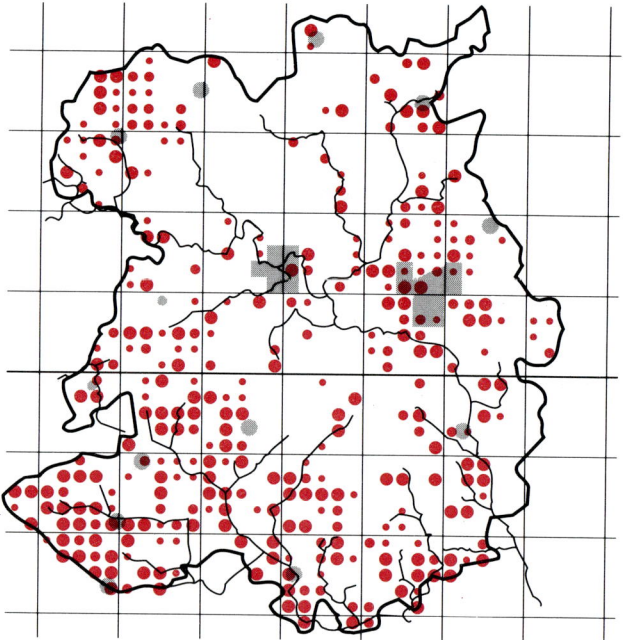

This delightful wagtail is closely associated with well-oxygenated, fast-flowing rivers and waterways. The contrasting blue-grey upperparts and vivid yellow vent and rump, together with the incessantly bobbing tail, brighten even the most picturesque of rivers.

Distribution is centred on the south-west uplands, particularly the tributaries of the Clun, Teme and Onny, and the Oswestry uplands, the Clee hills and the streams flowing into the Severn in the Telford region. However, most of the major rivers of the north, and the Severn below Bridgnorth, also hold some breeding Grey Wagtails. Even in these lowland situations they are closely associated with stretches of "riffly" water such as weirs, river confluences, or outfalls. Canals too may be used, favoured locations being not too distant from locks or feeder channels where some fast-moving water occurs.

Breeding sometimes begins early, in March, but most incubation is from mid-April through May, with repeat layings after failed clutches. Two, rarely three, broods may be reared. Nest sites, on rock or wall ledges or under tree roots, are invariably above flowing water. Therefore nests are difficult to find and proving breeding at the chick or nestling stage is best obtained by waiting for adults to return with food. Successful pairs are more easily confirmed when the chicks have fledged. The high percentage of possible breeding may be due to unsuccessful pairs being missed during incubation, foraging within adjacent tetrads, or post-breeding dispersal.

There is no evidence to suggest that breeding Grey Wagtails are anything other than resident. They are partial migrants in Britain, but the extent of movement through Shropshire is unknown.

Comparison with the BTO *Atlas* (1976) reveals confirmed breeding in 13 previously unoccupied 10-km squares, including some not wholly in the county, mainly along the southern edge of the north Shropshire plain. Some of these 10-km squares now have several occupied tetrads, so range expansion seems to have occurred since 1968–72, contrary to the national trend (*Pop. Trends* 1990).

A population of around 1–2 pairs per occupied tetrad, or 250–500 pairs in total, is estimated.

Ormerod *et al.* (1988) demonstrated a preference for basic streams rich in calcium salts which provide more abundant invertebrate prey than acidic waters, so afforestation of the south-west uplands and consequent acidification of run-off water may perhaps have affected Grey Wagtail abundance by reducing the density and diversity of this prey. However, *Pop. Trends* (1990) states that deciduous trees lining the river help to counteract the effect of water acidity, presumably by providing more non-aquatic insect prey, making Grey Wagtail less vulnerable than Dipper in the same upland habitat. *GT*

PIED WAGTAIL
Motacilla alba

Status: **Resident**

Tetrads with evidence of breeding

Confirmed	518 (60%)
Probable	150 (17%)
Possible	129 (15%)
Total	**797 (92%)**

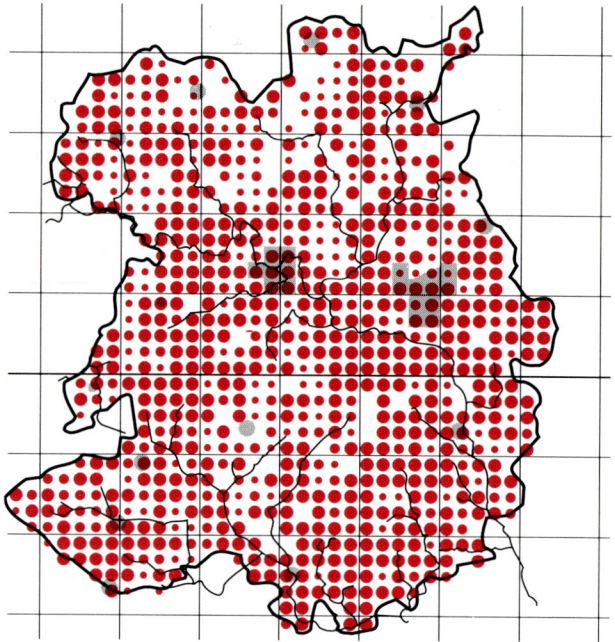

Pied Wagtails are seen throughout the year, but in winter some local birds migrate to southern Europe, especially in really hard weather. The whole breeding population of northern Britain moves south in the autumn, resulting in large migratory flocks and winter roosts in the Severn valley.

In April the winter flocks disperse to the widespread breeding areas, from town centres and farms to upland valleys and the rocks on both summits of Brown Clee. The gaps mainly correspond to woodland and moorland, which are avoided; and well-drained cultivated areas with no substantial streams (Map 2).

Nests are built in cavities and recesses. Stone walls, especially farm buildings, and banks and rocks are the main sites. An old House Martin nest on Atcham bridge was used in 1985 (*SBR*). Flat open areas, where small insects can be caught, are essential. Those frequented include lawns and playing fields, farmyards, well-grazed pasture, sewage farms, the banks of pools and streams, and even roads.

Pied Wagtails are relatively easy to locate. Contrasting black-and-white plumage, bounding flight, prominent perching, flicking tail and territorial call are all distinctive. Mixed farms are a favourite habitat, and cow sheds are especially favoured. Here breeding is easy to confirm, as both parents alight on the roofs while carrying food to the nest. Fledglings are also conspicuous, as they gather on the same roofs, or open ground, to be fed. Two or three broods are usually raised. However, most of the unconfirmed records, and a few of the gaps, will relate to isolated pairs in tetrads with little suitable habitat.

The Pied Wagtail is more catholic in its choice of nesting habitat than the other two wagtails, and is more numerous, though the population will be reduced if the preceding winter is harsh. Atlas fieldwork suggests an average of 3–6 pairs per tetrad, which is consistent with the national average (BTO *Atlas* 1976), giving an estimated population of 2500–5000 pairs.   *LS*

DIPPER

Cinclus cinclus

Status: **Resident**

Tetrads with evidence of breeding

Confirmed	132 (15%)
Probable	34 (4%)
Possible	36 (4%)
Total	**202 (23%)**

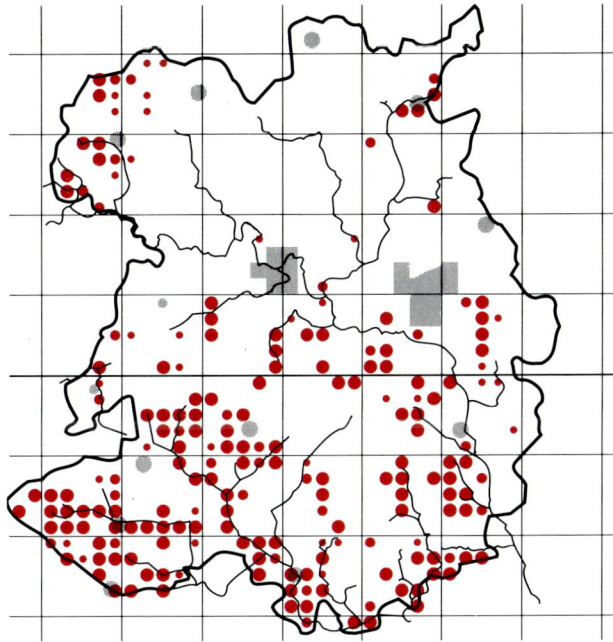

Dippers remain at their breeding sites throughout the year. Nest-building begins early in the season and some nests may contain eggs by the end of February. Two broods are usual, with some late nests still containing young in July.

A quick look under a bridge, which is a favourite site, is often all that is required to confirm breeding. The large domed nests, made of moss, are usually obvious, positioned on top of girders, pipes or ledges. Those in natural sites can be exceedingly well hidden under banks or large rocks, or deep amongst tree roots, and here confirmation of breeding can be obtained fairly easily by watching adults back to the nest when carrying food.

Dippers are usually restricted to fast-flowing, shallow rivers and streams where they obtain all their food. The stronghold is in the south-western hills, mainly along the rivers Onny, Clun, Teme, Rea and their tributaries, including the upper reaches of The Long Mynd valleys. Smaller numbers are found in the Oswestry uplands, on the upper reaches of the River Tern, along the Rea, Cound, Harley, Farley and Borle Brooks, on the rivers Worfe and Corve, and occasionally elsewhere. The plains are unsuitable as watercourses are too deep and slow-moving. Isolated records in unsuitable areas probably refer to juveniles dispersing from their natal areas.

During the breeding season Dippers are highly territorial, with lengths of 0.5–1.0 km usual for rivers of similar morphology and water quality to those found locally (*BWP*). Most occupied tetrads would therefore contain only 1–3 pairs, giving a population of 160–480 pairs.

Recent research suggests that the numbers of Dippers present at communal bridge roosts during the late autumn and winter is a good indication of the level of the local breeding population the following spring (Tyler *et al.* 1990; Ormerod & Tyler 1990). Over a wide area of mid- and south Shropshire numbers declined markedly during the last two years of the Atlas period. Counts were made each winter during the period 1987–91 at 30 different roost sites covering the whole of the stronghold and the total number of individuals fell by over 50%. Counts were as follows: 1987–88, 86; 1988–89, 85; 1989–90, 55; 1990–91, 39 (A.V. Cross, unpub.). This fall was mirrored in nearby areas of Powys (S.J. Tyler, pers comm.) and almost certainly results from the two very dry summers of 1989 and 1990 when stream levels were extremely low. Under these conditions pollution and siltation levels increase, causing a drop in the abundance of oxygen-loving invertebrates such as mayflies, stoneflies and caddisflies on which Dippers feed. The low water level may also lead to increased rates of predation of both adults and nestlings. There is no evidence to suggest that water acidification, through either acid rain or afforestation, has had any wide detrimental effect on Dippers here.

They are able to recover rapidly from a low population level and sizeable fluctuations may not be unusual. Indeed, following a wet summer, roost counts in autumn 1991 suggest a slight increase with a total of 42 Dippers found at the same 30 sites.

AVC

WREN

Troglodytes troglodytes

Status: **Resident**

Tetrads with evidence of breeding

Confirmed	740 (85%)
Probable	102 (12%)
Possible	27 (3%)
Total	**869 (100%)**

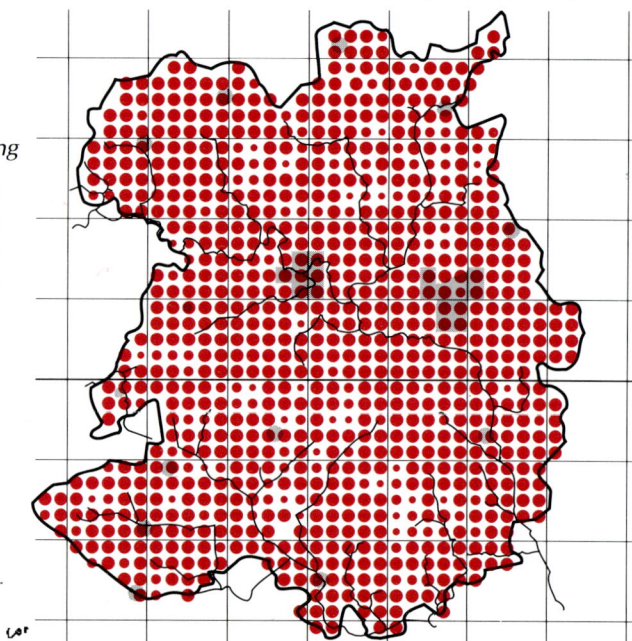

The Wren is a very common resident found throughout the county in an extremely wide range of habitats, especially in woodland, scrub, hedgerows and gardens. Well-grazed fields and some arable crops may not suit it but there will usually be a hedge-row or scrubby corner that will hold a pair or two. The valleys leading into the upland areas provide the routes for the Wren to reach the highest hills where it has been found nesting in the bracken. Nests are usually built in bushes or crevices, but one nested side by side with a Spotted Flycatcher on an old Swallow's nest near Chelmarsh and another occupied an old House Martin's nest at Atcham bridge (*SBR* 1989).

Its loud song makes location easy and, once the young have hatched, the adults can be seen carrying food and heard to protest vigorously when approached. Two broods are usual and newly fledged young call loudly, resulting in a high proportion of confirmed breeding records.

Hard winters are the main influence on the population level which fluctuates considerably. The *Handlist* (1964) commented that "very few survived the winter of 1962–63; it appeared to be the species worst affected by this winter" and numbers also fell, to a much lesser extent, in the hard winters of 1978–79, 1981–82, and 1985–86, with subsequent recovery. With the winters towards the end of the present study being relatively mild numbers are, once again, high. The BTO *Atlas* (1976) gave an average of 120 pairs per tetrad at a time when populations were also high. Assuming Shropshire is typical the population would be about 100,000 pairs.   CEW

DUNNOCK
Prunella modularis

Status: **Resident**

Tetrads with evidence of breeding

Confirmed	665 (76%)
Probable	139 (16%)
Possible	59 (7%)
Total	**863 (99%)**

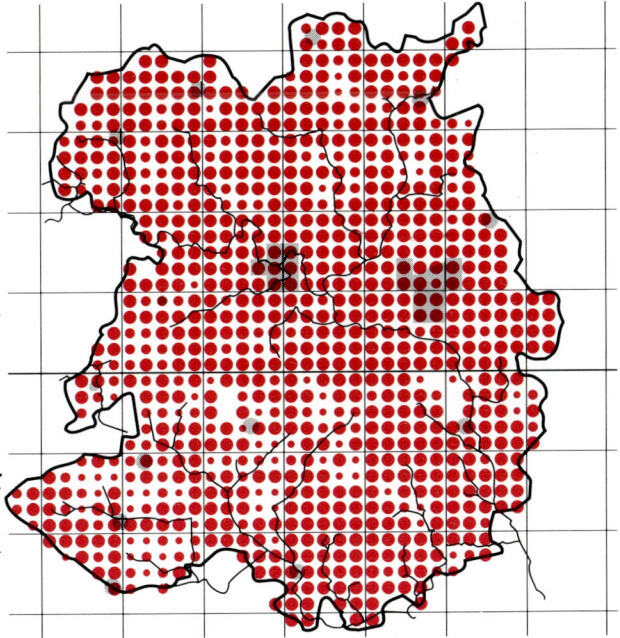

The Dunnock, or Hedge Sparrow, is common, yet unobtrusive and undemonstrative for most of the year. Inhabiting town and countryside, its only nesting requirement is thick low cover such as hedges, bushes, dense undergrowth or even bracken on the highest hillsides. Although tending to avoid closed-canopy woodlands, other than at the edges or in clearings, its catholic taste in habitat means there are breeding Dunnocks in every tetrad.

Most records are of confirmed breeding. Presence is betrayed by the thin but sweet warbling song, and although it is stealthy when feeding young and the nest is hard to find, patience is rewarded when adults are seen carrying food. Fledglings are usually more confiding, and two broods are usually raised, making confirmation easier. Pairs fostering Cuckoos also confirm breeding for both species.

Nationally the population is decreasing slowly for reasons that remain obscure. Dunnocks have a far-from-straightforward breeding pattern. Both sexes defend separate territories early in spring, and then combine these where possible as the season gets under way. A survey conducted at Nedge Hill, Telford (Bishton *SBR*

1983) and other studies further afield found polygamous behaviour with a range of male : female breeding ratios (*Pop. Trends* 1990).

Population estimates are therefore best expressed in terms of a number of territories. The national estimate is 2 million or an average of about 750 per occupied 10-km square. As western England holds a relatively low density (*Pop. Trends* 1990) but there are no large unsuitable areas in Shropshire, the figure may be around 25–30 per tetrad or 22–26,000.   DS

ROBIN
Erithacus rubecula

Status: **Resident**

Tetrads with evidence of breeding

Confirmed	794 (	91%)
Probable	63 (	7%)
Possible	12 (	1%)
Total	**869 (**	**100%)**

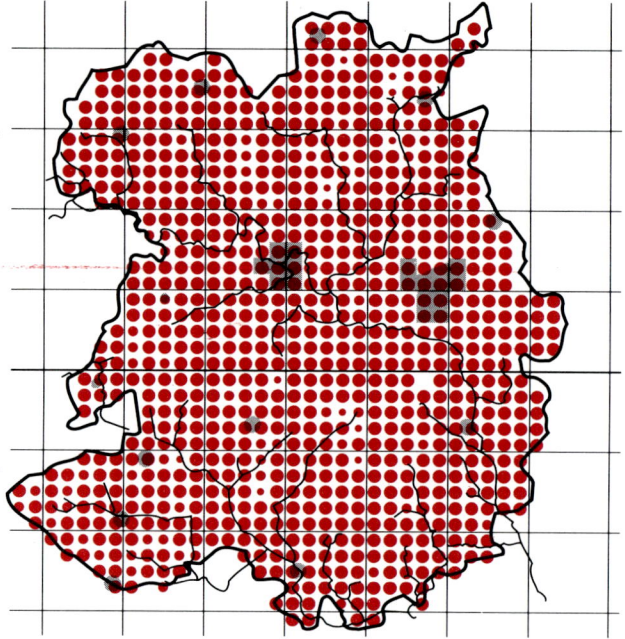

The Robin is probably the best known and most popular of our resident birds. Originally a woodland species it has adapted well to new conditions created by man.

Unusually, both sexes sing and defend ranges throughout the winter. After pairing males continue these duties and territories are easily located. Odd nest sites in sheds or outbuildings often feature in the local press. More regularly the nest, concealed in dense cover, is difficult to find. Two or three broods are raised and confirmation of breeding is much easier when adults can be seen carrying food to the nest, or when young have fledged. The high-pitched alarm call of the adults is a good indicator that young are present.

The Robin is one of our most common and widespread birds. Breeding densities are highest in woodland and mature gardens, with fewer pairs in intensively cultivated areas and on high ground. Numbers are greatly reduced following severe winters such as 1981–82, when they may disappear from marginal sites. These are later recolonised after numbers have built up in favoured areas. Following several relatively mild winters Robins have increased steadily, and in 1989 they reached their highest level on a local CBC plot since recording began in the mid-1960s (T.W. Edwards, pers. comm.). Assuming an average of 30 pairs/sq. km for woods and farmland, as suggested by CBC fieldwork (BTO *Atlas* 1976), the population is probably near 100,000 pairs.

APD

NIGHTINGALE
Luscinia megarhynchos

Status: **Occasional**

Tetrads with evidence of breeding

Confirmed	1 (	0%)
Probable	1 (	0%)
Possible	8 (	1%)
Total	**10 (**	**1%)**

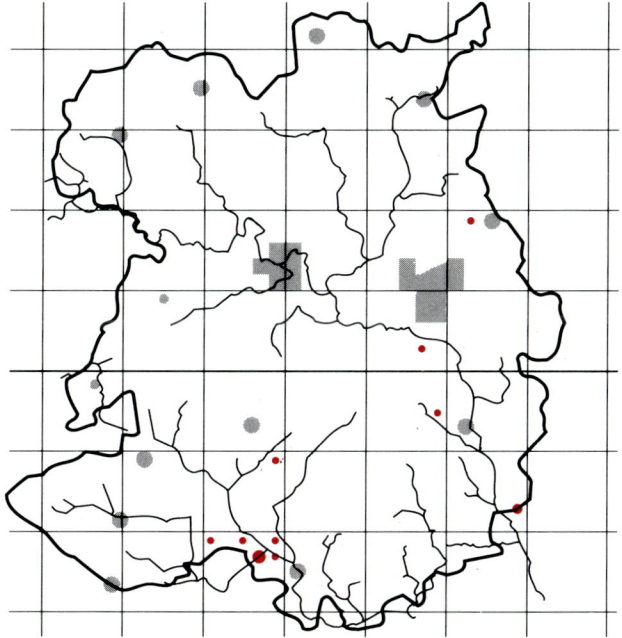

Though now very scarce, the *Handlist* (1964) recorded that before 1950 the Nightingale bred regularly in the Severn valley from Shrewsbury down to the Wyre Forest, with occasional pairs elsewhere. At the main colony, between Ironbridge and Coalport, a decline from up to 20 singing males in the 1940s to only 5 or 6 in 1963 was noted. Following a reported increase, a countywide survey in 1973 located 12 singing males, all in the Ironbridge area (Sankey & Wright *SBR*). This was a temporary fluctuation as, despite positive management of Lloyds Coppice, numbers continued to decline and the most recent record from this area was in 1980. Scattered records have come from elsewhere and numbers in the early 1980s ranged from 0 in 1984 to 6 in 1980 (*SBRs*).

The Shropshire breeding population was at the north-western end of an extension of the main range which followed the Severn valley (BTO *Atlas* 1976). Nationally the range has been contracting since at least 1950, resulting in a reduced population in the north and west. Climatic factors are believed to be responsible, rather than habitat loss (*Pop. Trends* 1990).

Nightingales typically inhabit thickets in woods, coppices and plantations with a rich shrub layer. Though the renowned song carries far it is heard less frequently during the day so they are hard to locate, and the extremely skulking behaviour makes proof of breeding or even presence of a pair difficult to obtain. The Atlas records were spread across the whole period, and most occurred in only one year. All were of singing males and, although they may not have attracted mates, one was heard long enough to qualify as holding territory. In one other case breeding was established when, in 1990, young were seen in a small wood near the River Onny following a number of records from this area in 1988 and 1989. The 1989 record from Benthall Edge comes from an area close to Ironbridge where coppice woodland, a favourite habitat, has been re-established.

So long as the present trend towards cooler, later springs continues, regular breeding in the Ironbridge area or elsewhere seems unlikely.   *CEW*

BLACK REDSTART
Phoenicurus ochruros

Status: **Probable**

Tetrads with evidence of breeding

Confirmed	0 (0%)
Probable	2 (0%)
Possible	2 (0%)
Total	**4 (0%)**

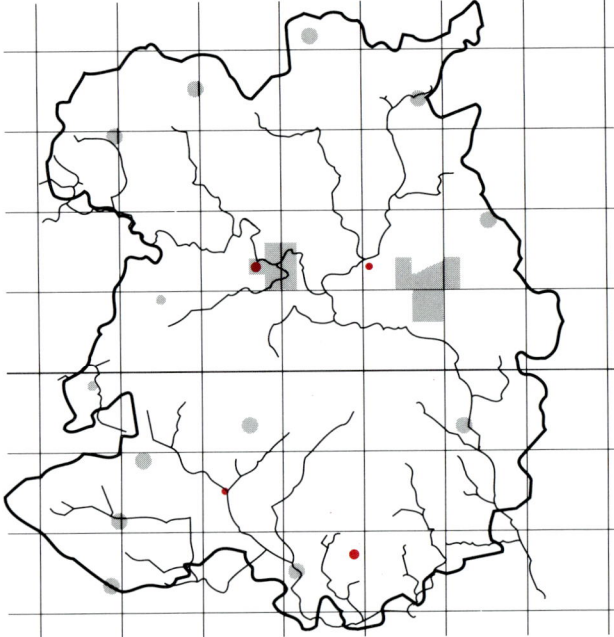

Looking like a melanistic version of his familiar cousin, the Black Redstart may occur as an irregular breeder, or as a scarce winter or passage migrant.

In Britain Black Redstarts breed mainly in power stations, other industrial sites, and derelict buildings. Unlike those on the Continent they do not use natural sites such as scree slopes or cliffs, although pioneer breeders in the 1920s used cliffs in Cornwall and Sussex. Many suitable local sites are inaccessible to birdwatchers, but such an unusual species, sporting a bright red tail, would attract the attention of workers and nesting activity should come to light. It is unlikely that breeding has occurred undetected.

Pairs were recorded at a Shrewsbury hospital and Titterstone Clee hill, with individuals at Craven Arms and Allscott. All these records are likely to relate to passage migrants.

Since 1956 there have been ten breeding season and ten passage or wintering records (S*BR*s). Confirmed breeding was recorded in 1963 when two pairs raised six young (*Handlist* 1964) and in 1978 when a pair raised four young (*SBR* 1978–79). All nests were in the Ironbridge area. A scarce migrant on the limit of its breeding range in Britain, local status is unlikely to change unless there is an expansion of the general range.   *JS*

REDSTART
Phoenicurus phoenicurus

Status: **Summer visitor**

Tetrads with evidence of breeding

Confirmed	196	(23%)
Probable	70	(8%)
Possible	58	(7%)
Total	**324**	**(37%)**

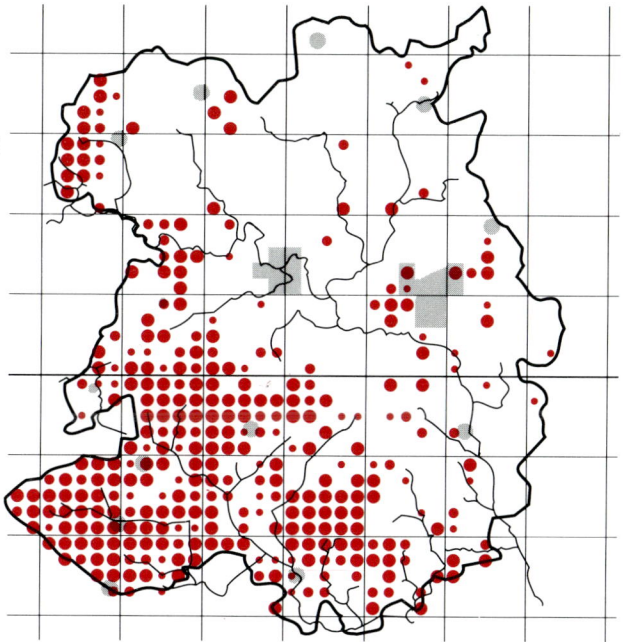

The male Redstart is one of the most colourful of our summer visitors. Singing begins as soon as they return from West Africa and by early May territories can be located. Confirmation of breeding can be obtained later as adults collect insects and larvae for their nestlings. Holes in trees or walls provide most nest sites, though occasionally nestboxes may be used. Recently fledged young are easy to locate, as they fly from one bush to another when approached. Once independent they wander extensively before journeying south, and some records away from the main breeding areas could refer to these juveniles, or adults on passage.

Redstarts occur mainly where open deciduous woodland, mature hawthorn hedges and scattered trees still remain on the marginal farmland of the Clee hills, The Long Mynd, Stiperstones, Clun Forest, Oswestry uplands, Middletown Hill and Loton Park. A few isolated populations exist in the north and east where similar habitat remains, the main pockets being found on Lilleshall Hill, The Wrekin and some of the other woods around Telford.

The population crashed in the early seventies, reaching a low in 1973, attributed to drought conditions in the Sahel where they winter. Since then numbers have increased slowly and by 1988–89 the pre-drought levels were exceeded (*Pop. Trends* 1990). During the Atlas period they extended their range in the north-west from the stronghold in mid-Wales, and occurred regularly on a local CBC plot for the first time, increasing to 3 pairs (T.W. Edwards, pers. comm.), reflecting the strong position which Redstarts are currently holding.

Twelve family parties were seen in one morning on the southern slopes of Brown Clee (*SBR* 1988), but tetrads at the edge of the range only support a few pairs. Atlas fieldwork suggests an average of 5–15 pairs in occupied tetrads, indicating a population of between 1400 and 4200 pairs. Suitable habitat bordering their existing range will allow for further expansion if conditions remain favourable. *APD*

WHINCHAT
Saxicola rubetra

Status: **Summer visitor**

Tetrads with evidence of breeding

Confirmed	35(	4%)
Probable	20 (	2%)
Possible	20 (	2%)
Total	**75 (**	**9%)**

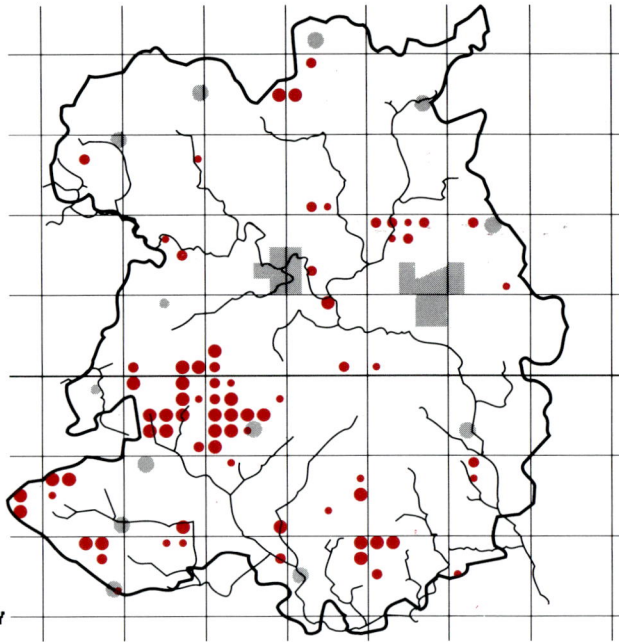

Whinchats arrive in strength in early May in their western stronghold — Bryn Shop, Rose Grove, Riddings, Mason's Bank, Black Bank and Cefn Hepreas in the Clun Forest; and Stapeley Hill, Heath Mynd, Norbury Hill, Black Rhadley Hill, Stiperstones, The Long Mynd, Earl's Hill and Caer Caradoc in the uplands — and also on Brown Clee, Titterstone Clee and Catherton Common. Thick bracken is the favoured habitat, usually on steep hillsides. Other tall herb plants and young conifer plantations are used by some pairs, and in the north they are found in the damp lowland mosses at Whixall Moss. Only two other isolated confirmed breeding records came from the lowlands, from near Culmington and near Brompton.

Migrating individuals and pairs will account for many of the possible and probable breeding records, especially away from the strongholds, and the cluster of records from The Weald Moors may reflect a regular staging-post, though it is equally possible that they do breed there in small numbers in terrain that is difficult to survey.

Breeding is usually fairly easy to prove. The male perches in full view on a tree or fence, or even a plant slightly taller than the surrounding growth, to proclaim territory or scold intruders, and both parents do so before taking food to the nestlings. When fledglings have just left the nest the whole family "tac" loudly.

The Whinchat has declined over much of lowland England since the 1950s or earlier, due to loss of habitat — roadside verges, railway embankments, derelict land and rough farmland (*Pop. Trends* 1990). The *Handlist* (1964) described it as "well distributed . . . wherever suitable localities occur, though more common on the high ground". This is no longer true, as with few exceptions it is now restricted to the uplands, and the range has contracted south-westwards compared with that shown in the BTO *Atlas* (1976). Several breeding sites used since then (*SBRs*) are no longer occupied, and the long-standing CBC plot near Oswestry had 9 pairs in

1968, 6 in 1977, and only one when it was last occupied in 1982 (T.W. Edwards, pers. comm.). The current strong dependence on upland bracken is probably the result of destruction of other suitable lowland tall herbs and grasses described in *Losing Ground in Shropshire* (*SWT* 1989).

Clun Forest, The Long Mynd, Titterstone Clee and Stiperstones have a minimum total of 36 pairs (*SBR* 1989), which, coupled with experience of Atlas fieldwork, suggests an average of 2–5 pairs/occupied tetrad, giving a population estimate of only 110–275 pairs.

LS

STONECHAT
Saxicola torquata

Status: **Resident**

Tetrads with evidence of breeding

Confirmed	13 (1%)
Probable	10 (1%)
Possible	14 (2%)
Total	**37 (4%)**

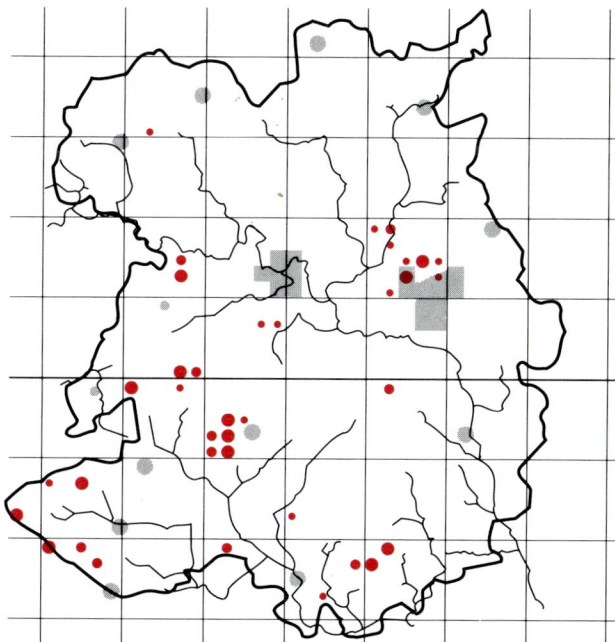

Stonechats usually inhabit the bracken-covered hillsides of The Long Mynd, Titterstone Clee, Clun Forest, the Stiperstones and Stapeley Hill. Slopes with gorse or other mixed vegetation, which provide prominent perches and cover for nests, are preferred. The closely related Whinchat is more numerous and widespread in these hills, presumably because it can utilise unbroken bracken.

Upland habitat is marginal for Stonechats, which are usually found on coastal and lowland heaths, so the breeding records from Catherton Common, Loton Park and north of Telford are perhaps more typical.

Males perch conspicuously, and call loudly. Once located, breeding is fairly easy to confirm as both parents carry food to prominent lookout posts before returning to the nest, fledglings are noisy when begging to be fed, and two or perhaps even three broods are raised. However, suitable habitat is often remote and difficult to survey, so the odd pair may have been overlooked.

As they wander widely in winter, any suitable area might be colonised. Hard winters reduce numbers considerably, and though isolated populations might be wiped out, multiple broods allow quick recovery. Distribution and population levels therefore fluctuate, but overall there has been a long-term national decline (*Pop.*

Trends 1990). Locally, Stonechats have "decreased considerably since the turn of the century", when Forrest considered them to be "common in summer on the moorlands"; they disappeared from regular pre-war sites in the Wyre Forest and at Llanymynech; and were "found breeding sparingly" in the western uplands with "occasionally an odd pair on Haughmond Hill" (*Handlist* 1964).

In apparent contradiction to this trend, current numbers and range are clearly much greater than reflected in the BTO *Atlas* (1976), which showed confirmed breeding in only one 10-km square in the south-west, and probable breeding in the two 10-km squares that include The Long Mynd and Brown Clee. Now breeding has been confirmed in nine 10-km squares, with probable breeding in two more, though not on Brown Clee. In the 13 intervening years sporadic breeding has been confirmed in several other widely scattered areas not occupied now (*SBRs*).

These fluctuations imply that the map overstates the distribution in any one year, particularly as wandering Stonechats will be responsible for some records. They are scarce, with perhaps only 1–2 pairs in each regularly used tetrad, suggesting a population of less than 25 pairs.   *LS*

WHEATEAR
Oenanthe oenanthe

Status: **Summer visitor**

Tetrads with evidence of breeding

Confirmed	43 (5%)
Probable	33 (4%)
Possible	27 (3%)
Total	**103 (12%)**

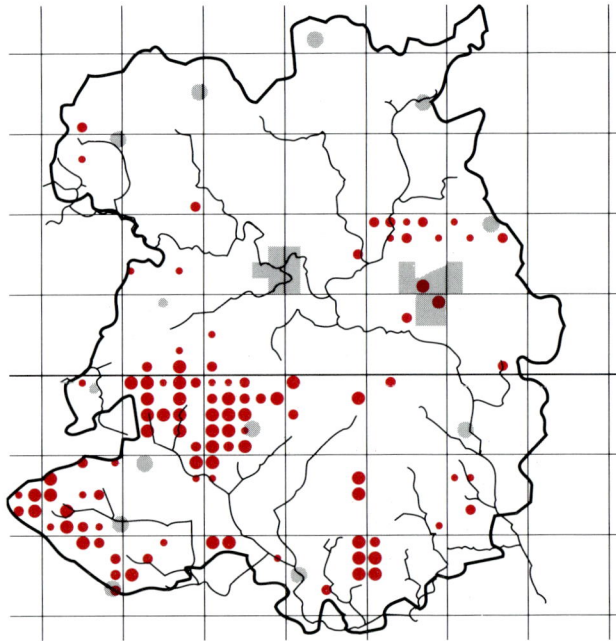

Wheatears usually return to their hillside breeding areas before the end of March. Short turf, created by grazing sheep or rabbits, or poor soil on steep rocky slopes, is essential for feeding; and nests are made in rock crevices, drystone walls or rabbit burrows.

Prominent perching on rocks or hillocks, a distinctive scratchy song and a spectacular display flight make Wheatears conspicuous while establishing territory. During incubation they are much harder to find, and several visits may be necessary to prove breeding, which is easiest when fledglings start to feed in the open, though adults may be seen carrying food to the nest hole earlier. Isolated pairs in remote areas may have been missed altogether, and some of the probable records in the south-west will almost certainly relate to nesting pairs.

The main concentration is on Titterstone Clee, where 26–40 pairs nest in the rocks on the summit and along Hoar Edge (SBRs). Rock habitat — drystone walls and old quarries and mineworkings — is also favoured by the few pairs on Brown Clee. Close-cropped grass with rabbit burrows is the main habitat elsewhere. Cefn Gunthly has several pairs, but the other hills of the west — Mucklewick Hill, Linley Hill, Black Rhadley Hill, Grit Hill and Stapeley Hill — have few. Black Mountain and other hills in the Clun Forest, the Stiperstones, each of The Long Mynd valleys, Ragleth Hill, Caer Caradoc and The Lawley also have few pairs.

Away from the south and west occasional pairs nest on the old pit mounds around Telford, but most of the singles and pairs will be migrants. The cluster of records on The Weald Moors now appear to relate to pairs in a favoured passage area.

The *Handlist* (1964) described Wheatears as "fairly common" in the Oswestry uplands, and added "a few pairs breed locally" north of Wellington. The BTO *Atlas* (1976) also had confirmed breeding records from north-western and central areas that are not occupied now. This indicates a decline over the last 20 years, certainly outside the main breeding stronghold. The population is known to fluctuate, but the reasons are unclear (*Pop. Trends* 1990).

Atlas fieldwork suggests an average of only 3–5 pairs in the 60 or so occupied tetrads in the south and south-west, giving an estimated population of 180–300 pairs.

LS

RING OUZEL
Turdus torquatus

Status: **Summer visitor**

Tetrads with evidence of breeding

Confirmed	2 (0%)	
Probable	4 (0%)	
Possible	6 (1%)	
Total	**12 (1%)**	

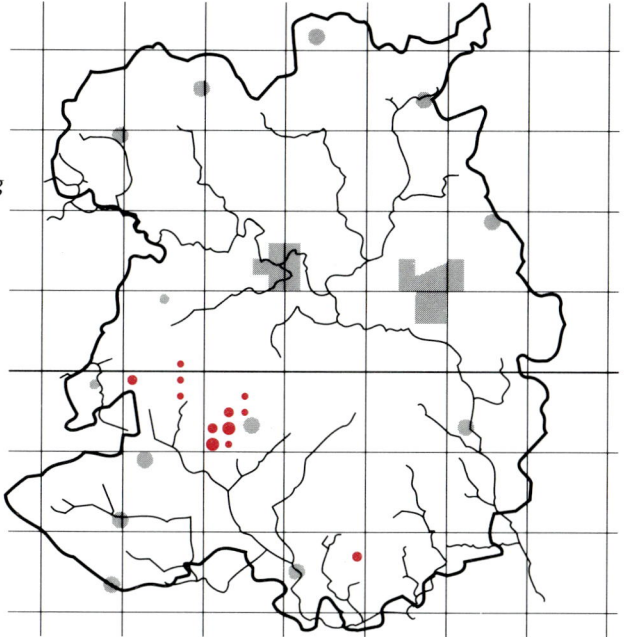

The Ring Ouzel is a summer visitor to hill and moorland, usually arriving from winter quarters in southern Europe or North Africa during April.

Highly specialised habitat is required: steep slopes with stunted trees or rock outcrops for song posts, short grass for feeding, and a stream. The appropriate landscape, unusual chattering call and prominent perches used for the far-carrying song all aid location, but it is very shy when approached. Breeding can be proved by watching adults taking food to the nest, or the heavily speckled fledglings. Two broods are usually raised.

Ring Ouzels breed only in the upper valleys of The Long Mynd. Although nationally the population now appears more or less stable, it did decline in the first half of the century, and numbers fluctuate considerably from year to year (*Pop. Trends* 1990). The *Handlist* (1964) referred to breeding on the Oswestry uplands, Clun Forest and Stow Hill "but not in recent years", and added "up to 10 pairs breed regularly on The Long Mynd, though numbers vary from year to year". This is little different from the current position: "at least 7 pairs of which 4 were confirmed breeding" (*SBR* 1987), and "probably a dozen pairs" (*SBR* 1989). Since 1964 all confirmed breeding records have come from The Long Mynd, except one from the Stiperstones in 1982 (*SBR*).

The probable breeding record for Brown Clee relates to a pair on 20 April and a single on 4 June 1987. The other records, including the pair on Stapeley Hill in April 1985, and a further two pairs on Brown Clee in April 1990 (*SBRs*), are almost certainly passage migrants.

In some parts of England numbers have fallen recently, probably due to increased human disturbance from hill walkers (*Pop. Trends* 1990), and this is a potential threat to The Long Mynd population.   *LS*

114

BLACKBIRD
Turdus merula

Status: **Resident**

Tetrads with evidence of breeding

Confirmed	832 (	96%)
Probable	34 (	4%)
Possible	4 (	0%)
Total	**870 (**	**100%)**

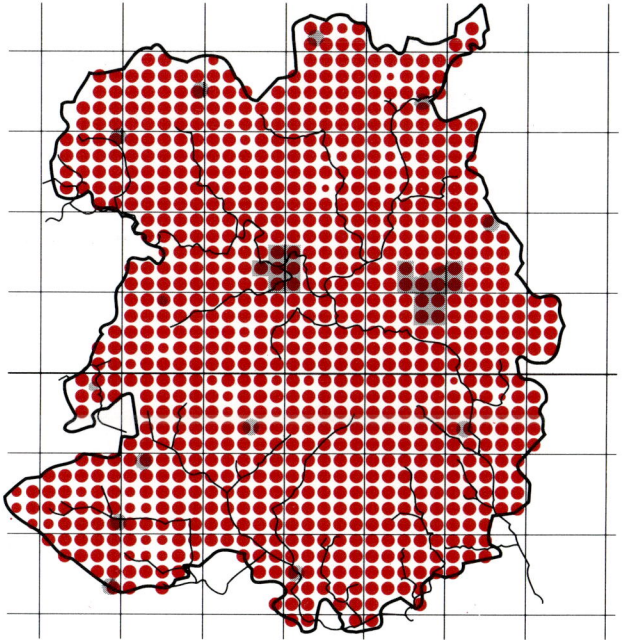

Perched on a prominent song post and in full flow from perhaps early February, this melodious songster is one of the commonest species.

The Blackbird is ubiquitous and no tetrad fails to offer suitable nest sites. It is easy to find as singing males advertise territories, and adults carrying food are conspicuous. Up to three broods may be raised in a season, so earlier failure to confirm breeding may be made good later.

The population density in favourable areas is very high and woodland edge, the ancestral habitat, is well represented, but tetrads containing large tracts of open country at high altitudes contain few pairs. Elsewhere in Britain surveys have shown wide differences in population: 35.7 pairs per sq. km on lowland farms, 66.2 pairs per sq. km in woodland and around 250 pairs per sq. km in suburbia (BTO *Atlas* 1976). Since then the removal of some hedgerows, severe trimming of others, and more intensive farming resulting in the destruction of scrubland and other suitable sites, have almost certainly reduced numbers. Also a slight national decrease has been recorded since 1976 on CBC plots, thought to be related to lower average winter temperatures, and this has been more noticeable on farmland than in woodland (*Pop. Trends* 1990).

Assuming the average density to be at the lower end of the range, as Shropshire is very rural with large areas of open country, an estimated 45–55 pairs per sq. km would give a population of 160,000–190,000 pairs.   *JS*

SONG THRUSH
Turdus philomelos

Status: **Resident**

Tetrads with evidence of breeding

Confirmed	572 (66%)
Probable	147 (17%)
Possible	107 (12%)
Total	**826 (95%)**

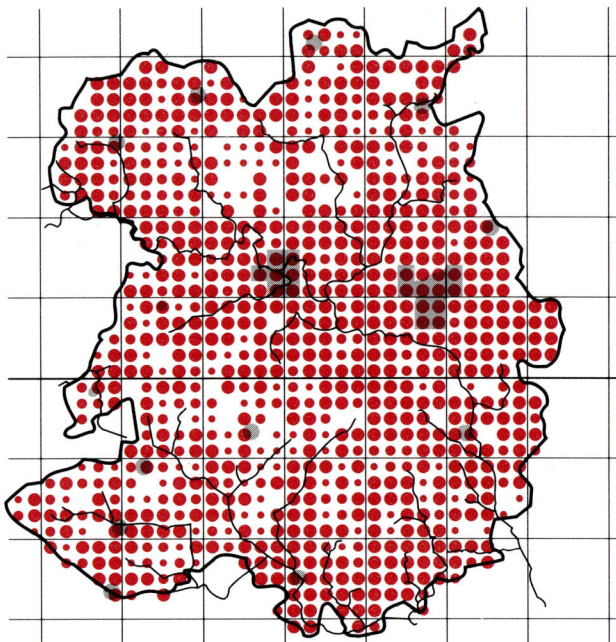

Delivered from a high perch, the clear repeated phrases of this major songster may be heard prior to the turn of the year, enabling territories to be located from early in the season.

Song Thrushes breed in any habitat with trees and bushes, especially woodland edges, farms, hedgerows, parks, large gardens and even modest suburban gardens. They are scarce or absent from both the uplands in the south and south-west, where bare areas and high ground are unsuitable, and those agricultural areas in the north where good quality hedgerows with song posts are lacking. Breeding is usually confirmed by seeing parents gathering food in short grass to feed nestlings, or recently fledged young, which may even be on the wing before hedgerows have broken into leaf.

Forrest (1899) described it as "plentiful everywhere" and in his day it far outnumbered the Blackbird. Change started during the 1920s and 1930s, when Blackbirds increased due to climatic amelioration, and became more marked from the 1940s when the Song Thrush went into decline.

The *Handlist* (1964) regarded it as still "common throughout the county, especially on low land", but continued: "The species suffered quite badly in the winter of 1962–63 and some birds moved considerable distances from the county". Cold weather affects the population adversely, and the consecutive severe winters of 1939–42 caused a setback from which it never fully recovered, with other severe spells to follow in 1946–47 and 1961–63. Apart from a few short-term fluctuations the decline has continued to the present day. However, since 1982 this has been steeper than expected from cold winters alone, and increased use of pesticides to kill slugs and snails may be having an impact (*Pop. Trends* 1990).

Numbers were very low after cold winters and wet springs in the early years of Atlas fieldwork, but increased somewhat over the later years. An informal enquiry by the County Bird Recorder among SOS members submitting regular records indicated "the species had definitely declined in recent years . . . but there appears to have been a shift in distribution over several years now, birds commonly being seen near human habitations . . . but becoming more scarce in open country" (*SBR* 1988). The BTO *Atlas* (1976) cited average figures of 13.5 pairs per sq. km on farmland and 27.1 in woodland, but the farmland population index has declined by well over half since then (*Pop. Trends* 1990). An average of 5–10 pairs per sq. km would give a population of 17,500–35,000 breeding pairs.   *JS*

MISTLE THRUSH
Turdus viscivorus

Status: **Resident**

Tetrads with evidence of breeding

Confirmed	524 (60%)
Probable	159 (18%)
Possible	101 (12%)
Total	**784 (90%)**

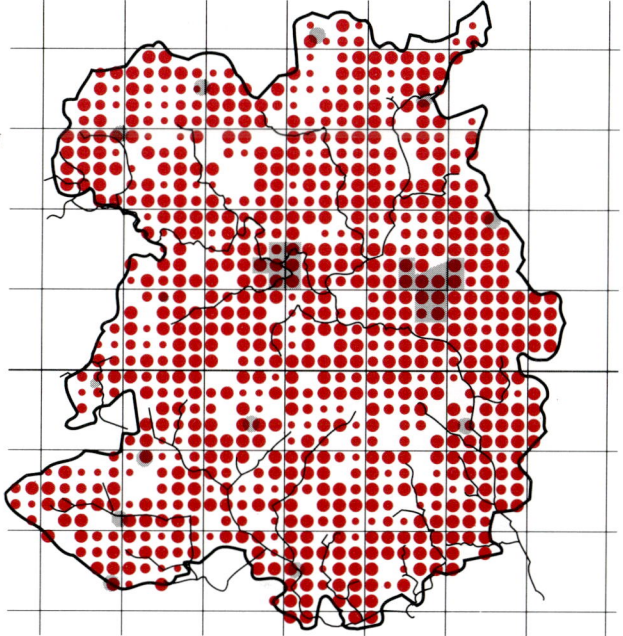

The habit of delivering his far-carrying, somewhat repetitive song from a prominent high perch, often just prior to the onset of rainy weather, gives the Mistle Thrush his provincial name of Storm Cock. It also makes initial location of a territory easy.

The *Handlist* (1964) reported them as "well distributed throughout the county on both high and low ground", and they still are, being found anywhere with good song posts, suitable trees for nesting and areas of short vegetation for feeding. They also inhabit well built-up areas provided there are large lawns, playing fields or parks, and they should be present in all tetrads apart from a few containing mainly woodland, open moorland, extensive arable areas, tall vegetation or parts of towns which lack large grassed areas. Territories are large, leading to a wide but thin distribution. In some of the apparently unoccupied tetrads they may have been overlooked due to confusion with the smaller Song Thrush, or low numbers in less favourable habitat. Breeding is usually confirmed by seeing adults gathering or carrying food, and is made easier as they forage up to a kilometre away from the nest, and raise two or even three broods.

Population densities vary according to farming practices. They probably benefit by a switch from hay to silage, but an increase in cereal production, particularly where autumn sowing reduces the amount of spring tillage, is detrimental (*Pop. Trends* 1990). The population is also adversely affected by severe winters and the *Handlist* (1964) stated: "Numbers were reduced following the winter of 1962–63". Cold periods in the early years of Atlas fieldwork will have affected them, and these followed other cold winters in the early 1980s, causing a fall in population, especially on farmland, since the most recent estimate of 300,000 pairs in Britain in 1982 (*Pop. Trends* 1990).

As Shropshire contains much suitable habitat, above-average figures of seven to eight pairs per occupied tetrad are likely, giving a population of 4800–5500 pairs.

JS

GRASSHOPPER WARBLER
Locustella naevia

Status: **Summer visitor**

Tetrads with evidence of breeding

Confirmed	8 (1%)
Probable	26 (3%)
Possible	53 (6%)
Total	**87 (10%)**

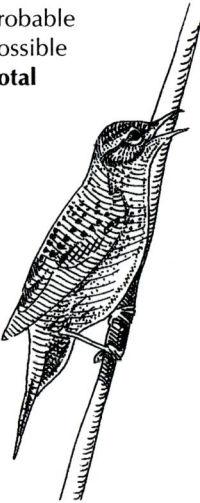

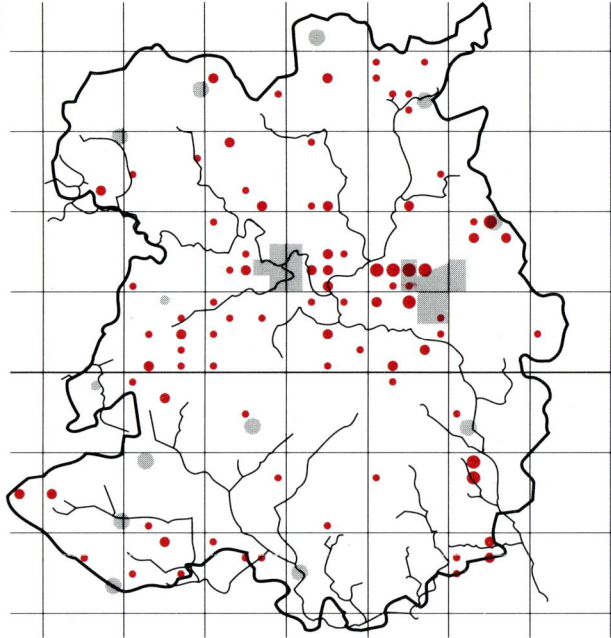

The Grasshopper Warbler arrives from mid-April to set up territory in both wet and dry habitats with thick, low, tangled growth, and also in young conifer plantations. More often heard than seen, its skulking behaviour, even when feeding young, and a tendency to sing mainly at dawn and dusk, ensures that it is frequently overlooked, and breeding is hard to confirm. Most of the records came from the Severn valley and the northern plain, with all the confirmed breeding records from damp lowland areas. There is also a scatter of records along the Rea valley reaching up into the Stiperstones, and another along the Teme valley and up into the Clun Forest. Although many of the small dots may relate to passage migrants, some of them, together with the probable records, are likely to indicate breeding pairs.

Numbers fluctuate annually and not all sites will have been occupied every year. Most records refer to isolated singing males although there were five at Haughmond Hill and three at Chelmarsh in 1989 (*SBR*).

The BTO *Atlas* (1976) suggested 10 pairs per 10-km square but the national population has declined significantly since then (*Pop. Trends* 1990). This decrease is reflected locally as Grasshopper Warbler occupied 14 of the 19 10-km squares wholly in Shropshire in the BTO *Atlas* (1976) but occupies only 9 of those squares now, in spite of more thorough fieldwork. Assuming an average of 1–2 per tetrad recorded, and that under-recording balances out the passage records, the population will be between 90 and 180 pairs.

CEW

SEDGE WARBLER
Acrocephalus schoenobaenus

Status: **Summer visitor**

Tetrads with evidence of breeding

Confirmed	40 (5%)	
Probable	36 (4%)	
Possible	40 (5%)	
Total	**116 (13%)**	

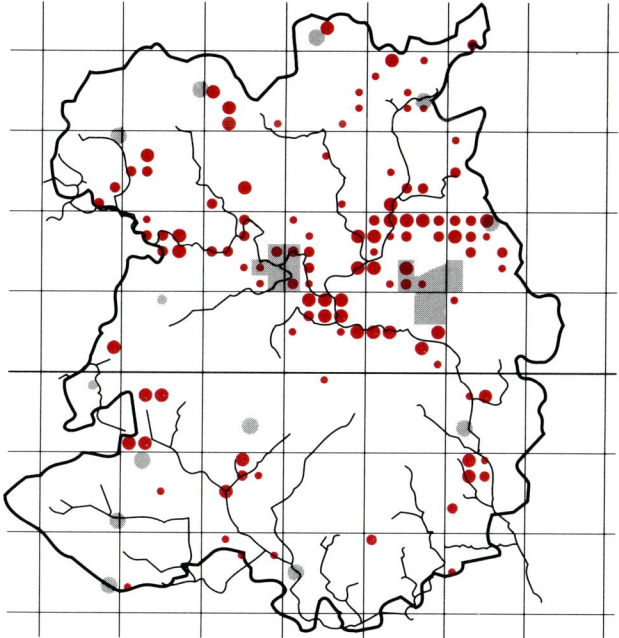

JPm'90

A summer visitor, the Sedge Warbler is found in rich vegetation alongside rivers and canals, around lakes, and in wet areas with bushes and scrub. Though often found with Reed Warblers in reed beds, it uses more varied habitats and is more widespread. Territories are easily located by the loud song of the male early in the season, and both adults may be seen later carrying food to the young. The Severn valley and the damp areas associated with The Weald Moors have produced most records, but there are some large gaps where it might have been expected, for example the lower Severn, the Rea valley west of Shrewsbury, and some of the slower rivers of the north. Although upland areas are generally unsuitable it does breed at Shelve Pool (300m) and The Bog (365m), close to the Stiperstones. Some of the small dots may represent passage migrants.

Land use close to rivers is important: where cattle have access to the water's edge the plants are often grazed and trampled but where arable crops border the river the bankside vegetation is richer and more suitable. Human activity such as fishing and boating can cause disturbance, and clearing access for fishermen sometimes destroys nests. On the Severn sudden increases in the water level following late spring and summer rain occasionally flood nests.

However, the major factor affecting population levels is the winter survival rate in the Sahel zone of Africa, which controls the numbers returning each spring to breed. Nationally the Sedge Warbler suffered a serious decline from the mid-1960s but there has been a substantial recovery since 1985 (*Pop. Trends* 1990). On a WBS site on 5.5 km of the Severn near Atcham the number of pairs has ranged from 20 in 1985 to 30 in 1990 with a peak of 33 pairs in 1987. However, another WBS site on the upper Tern near Market Drayton recorded Sedge Warbler only twice between 1983 and 1990, on passage. Only Allscott, Chelmarsh and Cranmere Bog are regularly reported as holding more than 3 pairs (*SBRs*) so, although tetrads with plenty of suitable habitat may hold up to 10 pairs, most hold less. Assuming an average of about 3–6 pairs per occupied tetrad, and adding a few for the possible breeding records, gives a population of between 250 and 500 pairs.   *CEW*

REED WARBLER
Acrocephalus scirpaceus

Status: **Summer visitor**

Tetrads with evidence of breeding

Confirmed	28 (3%)
Probable	14 (2%)
Possible	19 (2%)
Total	**61 (7%)**

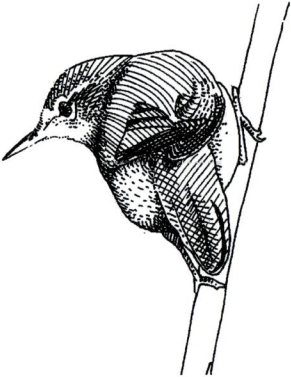

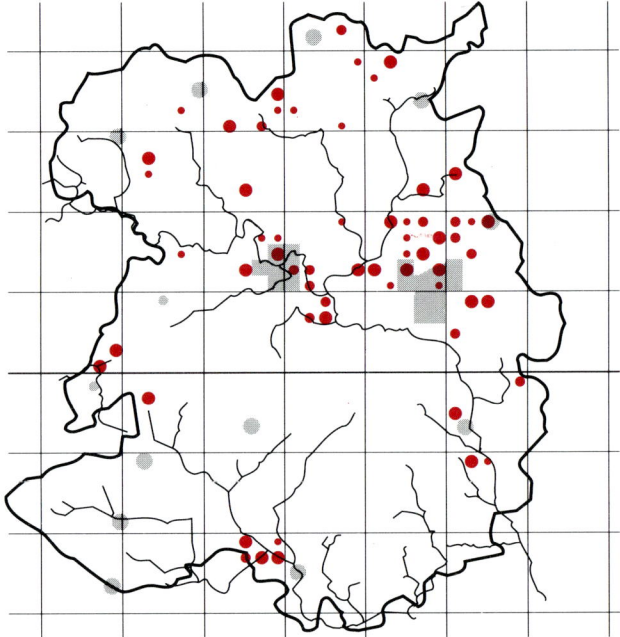

The Reed Warbler is a summer visitor mainly arriving in the first week of May although in some years a few arrive in late April (*SBRs*). It is closely associated with stands of the common reed, which grows at the margins of meres, pools, canals, and slow rivers in water up to 1.5m deep, and is locally frequent on the north Shropshire plain but rare elsewhere (*The Flora* 1985). The distribution map for common reed reflects that of the bird but there are tetrads with common reed but no records for Reed Warbler, especially in the north-west where the latter is at the western edge of its range in the Midlands.

Although few of the stands of common reed are extensive, quite small clumps can hold a colony. Reed Warblers will also nest in other vegetation close to the reed bed, and on the River Severn near Atcham two pairs regularly nested in a stand of reed canary grass where the common reed was absent. All sites are in the lowlands except Shelve Pool at 300m. The well-studied colony in a small reed bed at Allscott usually holds up to 40 pairs, and many adults return to this same area year after year, a few for as many as eight years and one for ten years (J.M. Langford, pers. comm.).

Protection of the reed bed habitat from drainage, clearance and scrub invasion is important if the Reed Warbler population is to be maintained. Numbers have increased at Chelmarsh following management work to extend the reed bed. Heavy summer rains can damage nests, and at Allscott in 1987 a summer roost of 10,000 Starlings had to be ousted from the reed bed to prevent total destruction of the colony. Losses at Fenemere have been blamed on predation by mink.

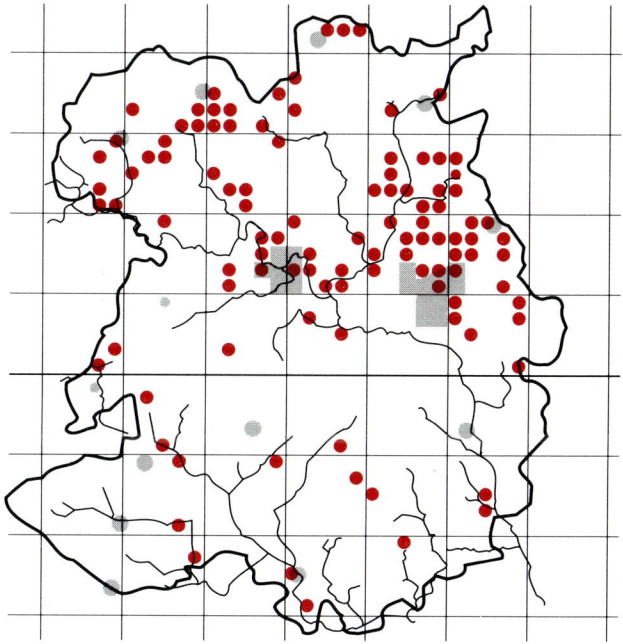

Distribution of Common Reed. (*The Flora*, 1985)

The loud song and specific habitat makes location easy and breeding can be confirmed by watching the adults flying to the reed bed with food for the young. Some of the small dots may represent passage migrants. Apart from the main sites at Allscott, Chelmarsh and Fenemere, most sites have few singing males (*SBR*s) suggesting an average of 5–10 pairs per occupied tetrad, so the population could be between 250 and 500 pairs.

CEW

JS

LESSER WHITETHROAT
Sylvia curruca

Status: **Summer visitor**

Tetrads with evidence of breeding

Confirmed	72 (8%)
Probable	112 (13%)
Possible	136 (16%)
Total	**320 (37%)**

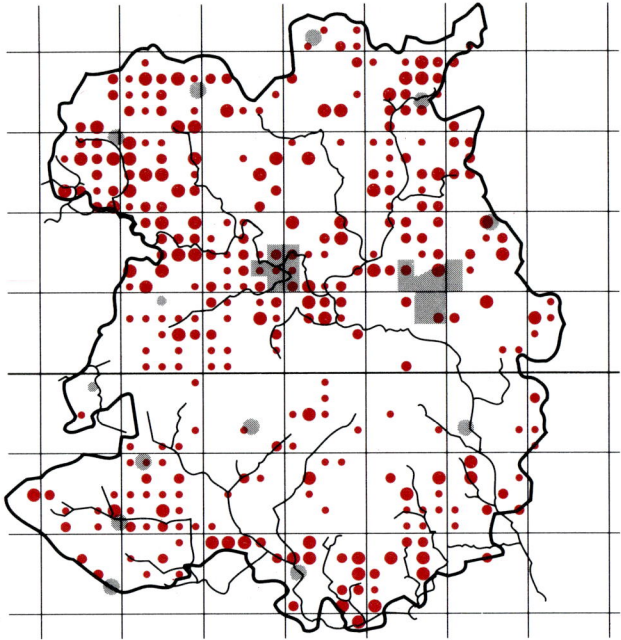

Males arrive mainly during late April, start singing immediately, and are frequently heard before being seen. The females follow a week or so later, pair bonds are quickly established, and subsequently the song is heard much less often.

Though they will nest in young conifer plantations, the preferred habitat is overgrown thick hawthorn and blackthorn hedges, dense scrub, and brambles, often with tall trees which may be used as song posts. Normally one brood is reared, but if eggs are lost through predation a replacement clutch may be laid.

As Lesser Whitethroats usually remain well hidden, sing from deep cover, have a brief song period and no song flight, they are less readily observed than their close relative, the Whitethroat. Breeding is most likely to be established when parents call loudly while feeding progeny. If young are threatened, the female will perform an injury feigning display to distract the intruder.

Lesser Whitethroats are unevenly distributed over the mixed farming areas in the north and south but tend to be absent from many central and eastern areas where arable farming has removed suitable habitat. Though they rarely breed above the 200m contour line due to lack of suitable habitat (BTO *Atlas* 1976), one tetrad with confirmed breeding in the extreme south-west has no land below 305m, and five occupied tetrads (including one in the north-west) have no land below 183m (see Map 5). Some of the possible breeding records will relate to passage or unmated males, but as some observers are not familiar with the song, and Lesser White-throats tend to skulk in thick vegetation, they are almost certainly under-recorded.

Only 3 10-km squares are unoccupied now, compared with 15 in the BTO *Atlas* (1976), consistent with the national expansion of range and increase in population density since then. Recent CBC mean farmland densities have been around 0.8 pairs/sq. km, but those in western England are lower than in the stronghold in the south-east (*Pop. Trends* 1990). As Shropshire is on the edge of the range a lower density is likely, and an average of 2–3 pairs/tetrad suggests a population of 700–1000 pairs.

In autumn they migrate south-easterly across Europe to winter quarters in north-east Africa, so Lesser Whitethroats are not affected by droughts in the Sahel, but nationally the population has fluctuated irregularly over many years. In common with other warblers they are dependent on conditions along the migration routes and especially in the wintering areas, as well as upon the availability of suitable breeding habitat. *MW*

WHITETHROAT
Sylvia communis

Status: **Summer visitor**

Tetrads with evidence of breeding

Confirmed	374 (43%)
Probable	238 (27%)
Possible	139 (16%)
Total	**751 (86%)**

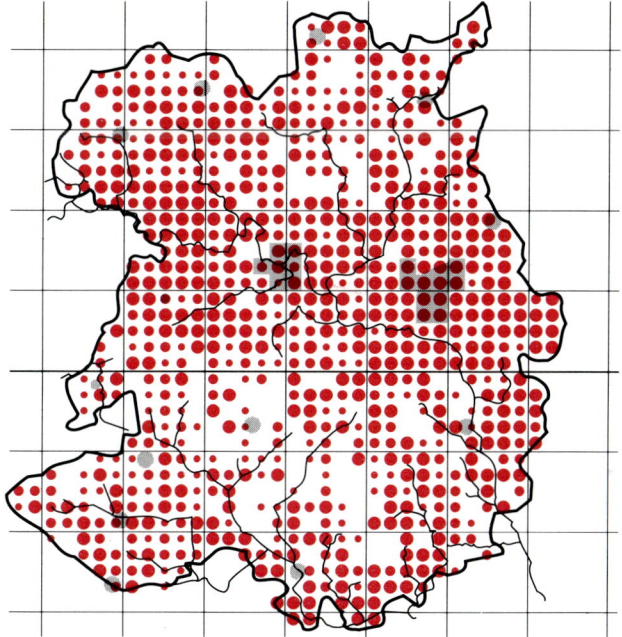

The first male Whitethroats usually arrive during mid-April, followed by females a fortnight or so later; by early May they are widespread. Less secretive than other warblers, they are sprightly and active, move easily among tangled growth, have a flitting, jerky flight and dart into the air to seize insects. Singing males will ascend up to 10 metres during frequent conspicuous display flights which aid location of territories. Courtship may continue throughout the breeding season as two broods are normally reared.

The favoured habitats are hedgerows alongside cultivated land and lanes; scrub and bushes, hawthorn thickets, and gorse on open ground; and woodland edge and young plantations. Breeding is readily confirmed when parents carry food to nestlings and fledglings, as the adults easily become very agitated and noisy at this stage, and the female will perform an injury-feigning display to distract intruders close to the nest.

Whitethroats occur more widely in the arable areas of the east and the mixed farming region of the north-west than elsewhere. They are also well distributed in the mixed farming districts of central and southern Shropshire but numbers are lower and distribution uneven due to restricted habitat. They tend to be absent from mature woodland and many areas of upland.

In 1985, when the Atlas started, the national CBC index was similar to that in 1972 when a population of 200 pairs/10-km square was estimated (BTO *Atlas* 1976) but numbers have increased since then, suggesting a population which fluctuates at around 7–15 pairs/tetrad, or between 5000 and 10,000.

Before 1968 this was the second most numerous warbler, described as "common throughout the county" in the *Handlist* (1964), but then a series of droughts seriously damaged the ecology of the wintering area in the Sahel south of the Sahara. The British population declined sharply and the CBC index fell to one-third of the 1968 level in 1969, and to only one-sixth by 1974. Locally the Whitethroat was described as "very common" in 1968, but only nine records were received in 1969, and in 1971 there were only two pairs on the long-standing CBC plot near Oswestry, compared with an average of seven before 1969. Numbers increased slowly until the early 1980s, after which another slump led to poor ringing totals in 1985 and 1987, followed by another increase towards the end of the Atlas period (*SBRs*). Nationally the CBC index is around one-quarter of the 1968 level and still fluctuates considerably (*Pop. Trends* 1990), so conditions in winter quarters determine the population level more than breeding success and habitats here.   MW

GARDEN WARBLER
Sylvia borin

Status: **Summer visitor**

Tetrads with evidence of breeding

Confirmed	216 (25%)
Probable	245 (28%)
Possible	183 (21%)
Total	**644 (74%)**

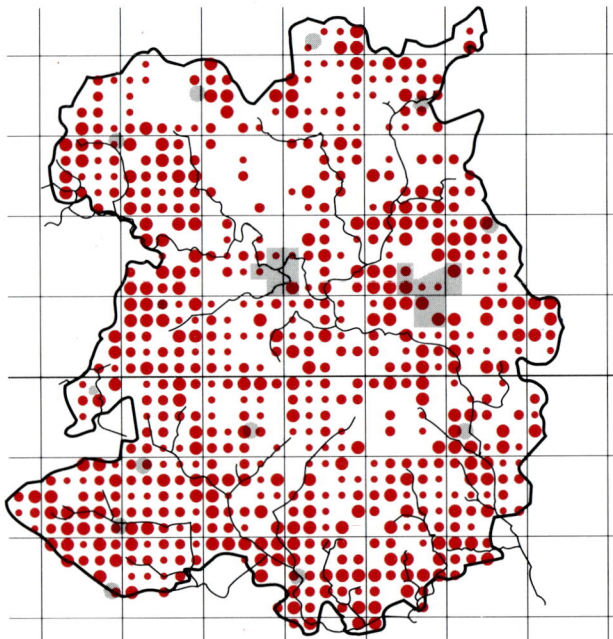

Male Garden Warblers arrive during April or early May, followed by females a week or so later. Though active, they tend to be secretive and quiet in movement, and remain hidden in foliage whilst searching for food. Presence is betrayed by the even-flowing sustained song, delivered from cover of trees or scrub but, unlike the Blackcap's, rarely from a prominent song post. As many observers have difficulty in distinguishing the song of Blackcap from that of Garden Warbler the latter may be under-recorded. For both species song continues from arrival until mid-July but is less frequent after pair-bond formation.

The Garden Warbler and Blackcap often nest in close proximity as their requirements are similar, though there is competition and the territories they defend do not overlap. The habitat preferred by the Garden Warbler is open deciduous woodland with thick undergrowth of tangled brambles, bushes and scrub; it will also nest in young conifer plantations, scrub without trees, dense overgrown hedges, or rank vegetation. The Blackcap favours woods and scrub with a higher, denser canopy which are more open and have less thick ground cover.

For both species probable breeding can be established provided visits are made before singing diminishes. Breeding is confirmed when parents carry food to young, though this is more difficult to observe for Garden Warbler which remains more under the cover of foliage. Alarm calls and injury-feigning displays are given by parents when young are approached or threatened by predators. The Garden Warbler usually raises a single brood whereas the Blackcap frequently raises a second.

Both species are well distributed but are more likely to be absent from central and east Shropshire, where many hedgerows and trees have been removed to facilitate modern arable farming techniques, so areas of scrub are few and suitable breeding habitat is limited. The Blackcap is far more numerous and has less exacting habitat requirements, and so is generally more widespread. However, the *Handlist* (1964) stated that the Garden Warbler predominated around Newport, Bucknell and Bishop's Castle. Though counts have not been made during fieldwork, comparison of the two maps shows 252 tetrads occupied by Blackcap but not by Garden Warbler and only 67 *vice versa,* but in the 110 tetrads south and west of SO39 the pattern is reversed with 15 tetrads occupied by Blackcap but not by Garden Warbler and 24 *vice versa*. The reasons for this are unknown but may perhaps be due to less intensive management of farmland and woodland, leaving more thick scrub and undergrowth, or more coniferous plantations, all providing habitats more suitable for Garden Warbler; or to the climate, as the spring may come too late on the higher ground for the earlier-arriving Blackcap, or the greater rainfall may affect breeding success.

In common with other trans-Saharan migrants the national Garden Warbler population declined from 1970 to 1976, when the Sahel drought was most severe, but has subsequently recovered. Thus conditions in wintering areas are important as well as those in the breeding habitat.

Nationally, the population appears to have increased by around 50% (*Pop. Trends* 1990) since the BTO *Atlas* (1976) estimated an average of 30–50 pairs per 10-km square which, if applied to Shropshire, gives a population of 1600 to 2600, or around 2.5–4 pairs in each tetrad where it was found. *MW*

BLACKCAP
Sylvia atricapilla

Status: **Summer visitor**

Tetrads with evidence of breeding

Confirmed	335 (39%)
Probable	311 (36%)
Possible	150 (17%)
Total	**796 (91%)**

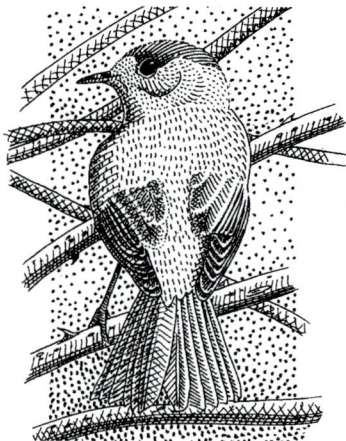

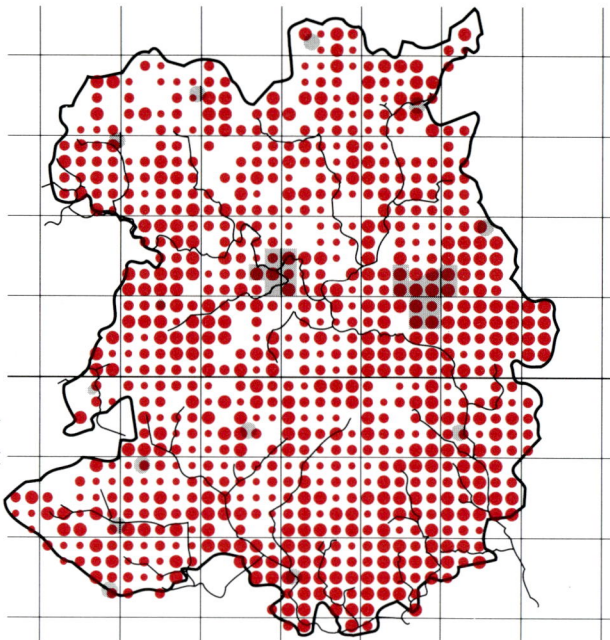

Blackcaps are present throughout the year and comprise two populations, a small one of Continental origin which over-winters, and the breeding population of summer migrants. Males of the latter group arrive from mid-April, followed by the females within a few days, and shortly afterwards they are widespread. One of Britain's best songsters, the rich, clarinet-toned song is delivered from a favoured song post or from tree tops or scrub.

Habitat and distribution are described in comparison with Garden Warbler previously. The Blackcap is well distributed throughout Shropshire except where there is an absence of suitable habitat in areas of large-scale arable farming, and in parts of the uplands in the west.

Unlike the Garden Warbler the majority of the British population appears to winter in the western Mediterranean basin and West Africa, so the Blackcap largely avoids the trans-Saharan migration to drought-stricken regions. The population therefore did not decline in the 1970s and has steadily increased in recent years for reasons that are unclear (*Pop. Trends* 1990). Nationally the population appears to be around four times that of the Garden Warbler, suggesting 6500–10,000, or 8–13 pairs in each tetrad where it was found.

MW

WOOD WARBLER
Phylloscopus sibilatrix

Status: **Summer visitor**

Tetrads with evidence of breeding

Confirmed	71 (8%)	
Probable	68 (8%)	
Possible	86 (10%)	
Total	**225 (26%)**	

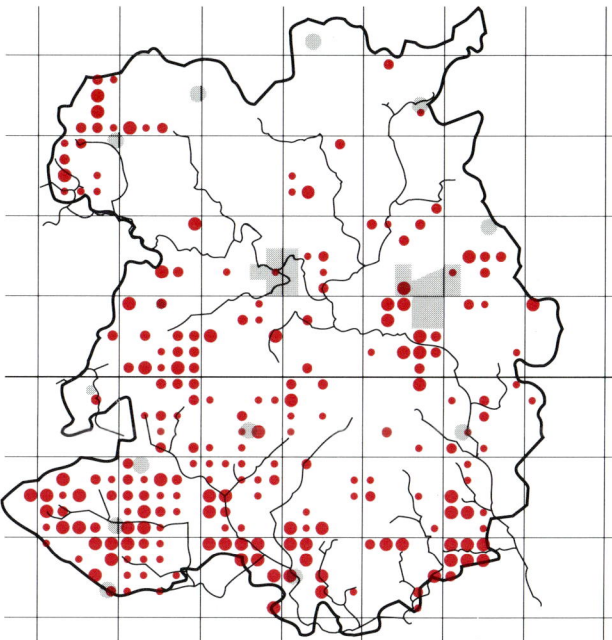

By late April the first male Wood Warblers have arrived, followed by females within a few days, and by early May they are widely but locally distributed. The song, a long quivering trill, has been the subject of poets' verse through the ages, and is mainly delivered during fine, still conditions, but less frequently in wind and poor weather, when presence is much harder to establish.

The Wood Warbler is far less common than the other leaf warblers, the Willow Warbler and Chiffchaff, due to exacting habitat requirements. Oak, beech or birch woodland with a high closed canopy is usually chosen, but it must be open beneath the canopy with only a scattered undergrowth of shrubs, brambles, bracken or grass. Woods with dense undergrowth are shunned and, unlike elsewhere in its range, conifers are not tolerated. Hillside sites with gradients from 10° to 75° are preferred, with the optimum being about 40°; flat areas are usually avoided (Bishton *SBR* 1983). The domed nest is built into the woodland floor, usually in a hollow.

The high proportion of possible breeding records arise from passage migrants, unpaired males wandering in search of a mate, post-breeding dispersal, and infrequent fieldwork visits to the more remote habitat. Proof of breeding is difficult to obtain because parents are frequently hidden by foliage when carrying food to progeny, the usual means of confirmation, though parental feeding does continue for about one month after fledging. Generally one brood is reared, though rarely there are two, for example in the Hope Valley (*SBR* 1986). Polygyny is common; a male will mate with a late-arriving female whilst the first hen is incubating eggs, and he is aggressive in defence of territory and will attack other leaf warblers.

The Wood Warbler is locally distributed due to its specialised habitat requirements, the main concentrations being on the western hillsides. Elsewhere suitable habitat occurs notably at Whitcliffe, Benthall Edge, The Wrekin, and the

Wyre Forest. The *Handlist* (1964) stated that it was possibly as common as the Willow Warbler in the Wyre Forest, and that deciduous woodlands, especially oak, were being felled and replaced by conifers, so habitat was being restricted and consequently the population was decreasing. Some 10-km squares show confirmed breeding in the BTO *Atlas* (1976) but not on this map, although generally the current results are slightly better. Wood Warbler numbers fluctuate from year to year, and it may desert woodland in which breeding has occurred in previous years.

Since unmated males also sing to defend territories, and polygamy occurs, the population is best expressed as territorial males rather than pairs. *SBR* records since 1983 list a total of 124 singing males at named locations covering up to 24 of the best tetrads. An average of 2–4 per occupied tetrad, plus 1–2 in tetrads with possible breeding, suggests a population of 375–750.

A systematic sample survey in 1984–85 to establish the British population found a mean of 13.8 male Wood Warblers per 10-km square where it was recorded in the Marches (Bibby 1989) which, after discounting part of the 10-km squares not wholly in Shropshire, suggests a population at the lower end of the above range, around 400 males.   *MW*

CHIFFCHAFF
Phylloscopus collybita

Status: **Summer visitor**

Tetrads with evidence of breeding

Confirmed	219 (25%)
Probable	361 (41%)
Possible	190 (22%)
Total	**770 (89%)**

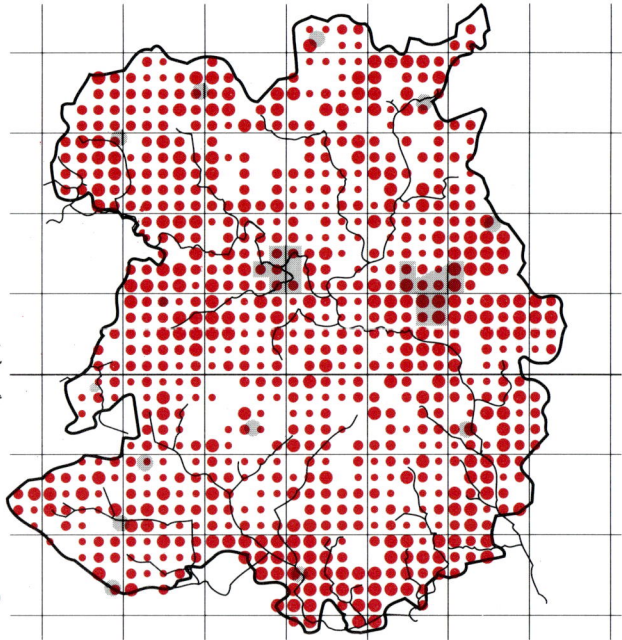

The Chiffchaff is amongst the earlier summer visitors, arriving from mid-March and widespread by the middle of April. It is readily distinguished from the closely related Willow Warbler by its simple disyllabic song, which is audible at 100 metres in still weather conditions and is often delivered from high branches in trees whilst flitting amongst foliage in search of food.

Primary habitat is mature deciduous woodland with undergrowth; also favoured are conifer woods, groves, shrubberies, overgrown hedgerows, and mature gardens in suburbs. The Chiffchaff usually avoids young plantations and scrub used by Willow Warblers because it is more dependent upon tall trees for song posts and for food in the canopy, whereas the latter forage amongst the scrub layer. One brood is generally reared, occasionally two. Almost half the records relate to probable breeding as the far-carrying song ensures that territories are easily located; fewer reports are of confirmed breeding as it is difficult to observe parents, often in undergrowth, carrying food to young, or separate them from the more common Willow Warbler when silent. As singing is reduced after pairing and observation in the canopy is difficult, many of the possible breeding records will relate to breeding pairs in tetrads visited infrequently, although some will arise from passage migrants.

The Chiffchaff is well distributed and only some arable farming districts and uplands lack suitable habitat. Atlas fieldwork suggests an average of 5–10 pairs per occupied tetrad which, after allowing for some of the possible breeding records relating to breeding pairs at lower density, suggests a population of 3000–6000.

The majority of Chiffchaffs winter south of the Sahara with the remainder in the Mediterranean basin or West Africa. During the last 30 years the British population has varied considerably because those making the trans-Saharan migration have suffered a high mortality in often drought-stricken regions. Those wintering further north have not endured such adverse conditions, thus avoiding greater losses. The population reached low points in 1976 and 1984 but has been increasing during the Atlas period (*Pop. Trends* 1990). In common with some other warblers, the fortunes of the Chiffchaff are dependent on circumstances in wintering areas as well as the availability of breeding habitat.   *MW*

WILLOW WARBLER
Phylloscopus trochilus

Status: **Summer visitor**

Tetrads with evidence of breeding

Confirmed	554 (64%)
Probable	218 (25%)
Possible	77 (9%)
Total	**849 (98%)**

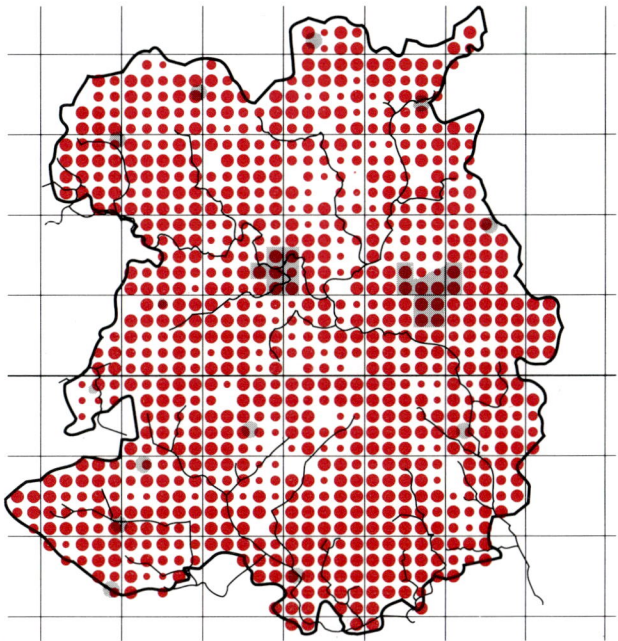

The Willow Warbler is our commonest summer visitor. After the appearance of the odd early arrival in late March the major influx usually occurs during the second week of April. The song, a fluent and wistful series of descending notes, advertises its presence in most areas where there are small trees or tall bushes. It tends to avoid woodland with a closed canopy, preferring clearings, rides and the margins of woods together with scrub, bracken-covered heaths, large gardens and even hedgerows.

Identification is based on song or call, since in appearance it is almost indistinguishable from the closely related Chiffchaff, which shares much of its habitat. Although nests are hard to find the song draws attention to the breeding territory, in which the carrying of nest-building material or food for young can be witnessed; the tendency to raise two broods is a further aid to fieldwork. The map demonstrates the wide availability of suitable habitat and the relative ease of confirming breeding. Tetrads of monoculture arable farming may contain little suitable habitat, and Willow Warbler may be absent from them or be so scarce as to be overlooked.

On the basis of a British population estimated at 2.5 million pairs, or an average of about 1150 pairs per occupied 10-km square, with western England having an above-average relative density (*Pop. Trends* 1990), the local figure is likely to be around 50 pairs per occupied tetrad, a total of at least 40,000 pairs.   DS

GOLDCREST
Regulus regulus

Status: **Resident**

Tetrads with evidence of breeding

Confirmed	177 (20%)
Probable	204 (23%)
Possible	172 (20%)
Total	**553 (64%)**

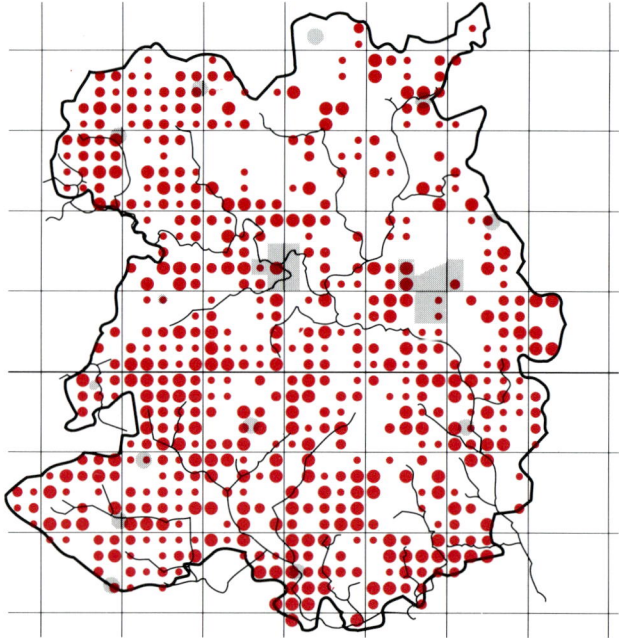

Planted coniferous woodlands are notorious for their lack of variety of all forms of wildlife, including birds. However, the few species which are able to adapt to such a habitat benefit, presumably from the lack of competition. The Goldcrest vies with the Coal Tit as the dominant species in these managed woods and occasionally reaches a very high density indeed. However, it is also resident at much lower concentrations in deciduous woodland.

The distribution correlates well with that of woodland (Map 8), especially conifers, and the gaps in the north and east reflect the scarcity of woodlands there.

Afforestation was predicted by the *Handlist* (1964) to increase the number of Goldcrests in Shropshire, but subsequent evidence for this is anecdotal. Large and rapid fluctuations due to high winter mortality, followed by surprisingly fast recovery, often with three or more broods per year, are part and parcel of its population dynamics. A large national decrease of 60% was recorded in the CBC population index between the summers of 1985 and 1986 (*Pop. Trends* 1990), attributed to prolonged winter frosts. Recovery in 1986 and 1987 soon brought numbers back close to pre-1986 levels, so it is unlikely that the Atlas has been significantly affected by the population crash, though some under-recording may have occurred in tetrads visited only in the first two years.

Goldcrests build tiny cup nests at variable heights in trees or shrubs, preferentially selecting conifers where present. They usually gather food high in the tree tops and nests are difficult to find so most confirmed records are of newly fledged young being fed by adults. Only 30% of all records are confirmed breeding but the majority of the rest undoubtedly refer to nesting pairs.

In 1972 the CBC recorded an average density equivalent to 50 pairs per tetrad on plots mainly in deciduous woodland and over 1000 pairs per tetrad were found in special surveys of spruce plantations (BTO *Atlas* 1976) at a time when Goldcrest numbers were relatively high. Though the CBC woodland index in 1988 was around half the 1972 figure, Goldcrests were still recorded in over 550 tetrads and with around 4% of the county planted with conifers, the population must be greater than 30,000 pairs most years and is likely to be above 50,000 pairs in some.   *GT*

FIRECREST
Regulus ignicapillus

Status: **Probable**

Tetrads with evidence of breeding

Confirmed	0 (0%)
Probable	1 (0%)
Possible	2 (0%)
Total	**3 (0%)**

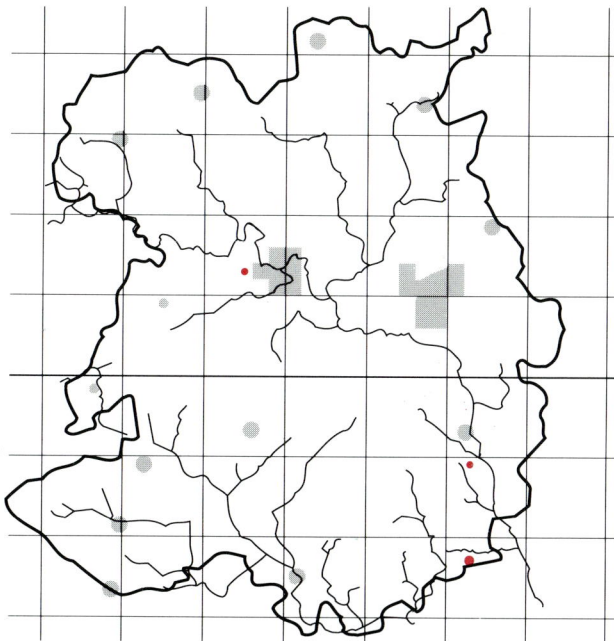

The Firecrest spread north-westwards across continental Europe, with the first confirmed breeding record in Britain in 1962 in the New Forest. Mainly a passage migrant and winter visitor, it is still a national breeding rarity and the annual number of recorded singing males fluctuated markedly between 29 and 102 during 1980–1987, apart from an exceptional 175 in 1983 (*Red Data* 1990). Breeding has been confirmed in the adjacent county of Worcestershire (BTO *Atlas* 1976).

In Shropshire the first *SBR* record was in 1972, but it has been recorded annually since 1974, except in 1982 and 1984, mostly between November and early April. Prior to the Atlas period there were only three breeding season records, all of individuals: near Chelmarsh in early April 1977; singing in Gatten wood, Stiperstones in May 1978; and feeding at Nedge Hill in May 1983. A juvenile found in Brown Clee woods towards the end of August 1979 may have been an indication of breeding nearby, though they wander widely from mid-July onwards.

Atlas records are of individual singing males at Mill Brook near Chelmarsh in May 1985, in the Wyre Forest in April and June 1989, and at Oxon Pool near Shrewsbury in April 1990. No females were seen and it is likely that all these records were of prospecting males overshooting, typical of many migratory species at the edge of their normal breeding range.

The Firecrest occurs in coniferous and deciduous woods but, in common with the Goldcrest, usually selects a conifer or evergreen such as ivy in which to build the tiny nest. Most observers are unfamiliar with the high-pitched song and call notes by which it is usually located and they are outside the hearing range of some, hence it may have been missed. Once incubation starts it becomes even more difficult to find as singing is then less frequent, but the scarcity of Atlas records probably does reflect the true status.

It may well be overlooked in the expanse of the Wyre Forest or in other southern woodlands but the tendency to breed in semi-colonial groups should ensure that any regular breeding is located. Isolated pairs may have gone unnoticed but the increasing frequency of winter records may well herald confirmed breeding in the not too distant future.   *GT*

SPOTTED FLYCATCHER
Muscicapa striata

Status: **Summer visitor**

Tetrads with evidence of breeding

Confirmed	525 (60%)
Probable	127 (15%)
Possible	96 (11%)
Total	**748 (86%)**

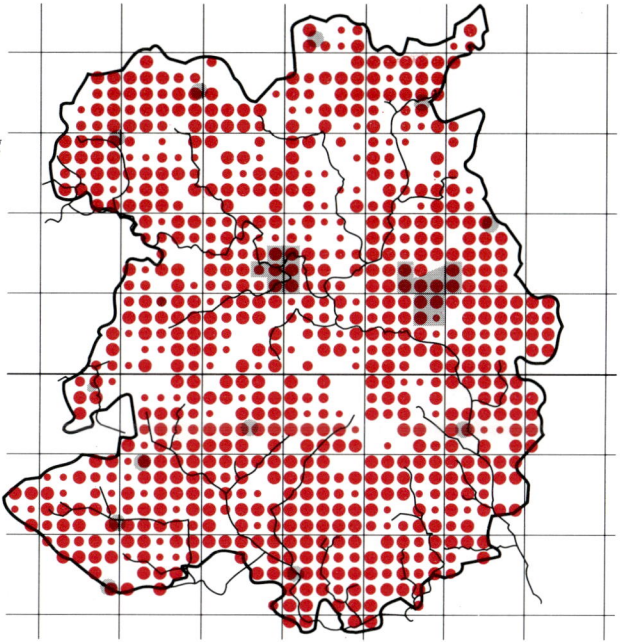

One of our latest arrivals, usually only appearing in strength towards the end of May, this summer migrant from Africa is found on the outskirts of woods and felled clearings, in hedgerows, parks and gardens, and around farmyards, pools and streams. The nest is made on a ledge, in a natural tree hollow or hole in a brick wall, or against a wall covered with creeper or shrub. An old House Martin's nest at Hopton Cangeford (*SBR* 1988) and an old Swallow's nest at Chelmarsh (*SBR* 1989) were used, the latter adjacent to a Wren taking the same opportunity. Successful nest sites are often used repeatedly, one porch in Prees sheltering generations for a final total of twenty-two consecutive years (*SBR* 1984).

The darting flight for flies, returning to the same perch, and a rather thin call-note, attract attention. Proof of breeding is usually straightforward; adults gather several flies before returning to the nest, and newly fledged young being fed are conspicuous, whilst nests made in gardens soon become known to human neighbours. In addition there may be two broods, especially in warm dry summers.

The gaps coincide with well-drained arable land, where copses and hedgerows may not provide sufficient flies for successful breeding, and the use of insecticides will further reduce prey numbers. When nesting away from human habitation or at low density it may have been overlooked, particularly during wet summers when both observer and flycatcher activity were relatively low.

Though the annual population fluctuates, there is a gradual, long-term decline and the biggest fall has occurred on farmland in southern Britain. The current national population is estimated at 200,000 pairs, or around 100 pairs per 10-km square. Since western England holds a relatively low density (*Pop. Trends* 1990) the corresponding county figure is likely to be an average of about 3 pairs per occupied tetrad, or around 2000 pairs. *DS*

PIED FLYCATCHER

Ficedula hypoleuca

Status: **Summer visitor**

Tetrads with evidence of breeding

Confirmed	154 (18%)
Probable	44 (5%)
Possible	61 (7%)
Total	**259 (30%)**

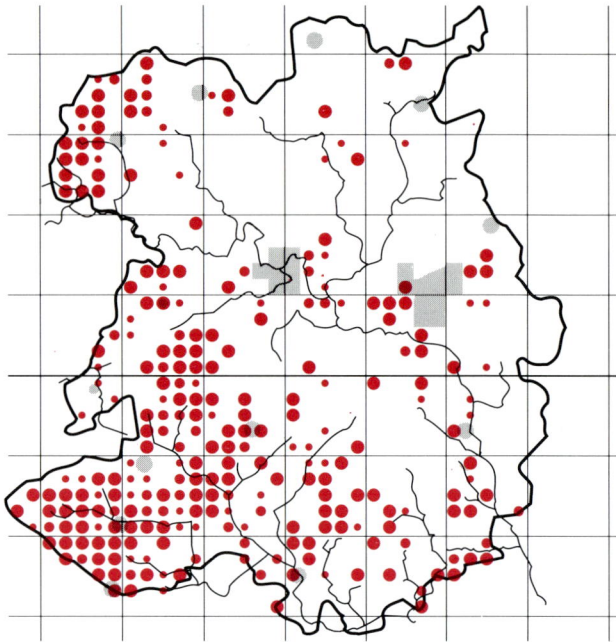

Arriving in mid-April, the adult male's distinctive pied plumage and persistent song soon attract attention. The preferred habitats are those open oakwoods which have little undergrowth, and stream-side alders, in the upland valleys of the west and south-west. Both adults will scold an intruder and are easily watched back to the nest hole, especially when feeding young. Some of the small dots will refer to passage migrants.

The *Handlist* (1964) noted that "it is found breeding regularly", mainly in the south and west, and had "increased its range considerably since H.E. Forrest wrote that it was an uncommon summer visitor (1899)". Apparently the Pied Flycatcher first reached Shropshire during an expansion of range in the 1880s and 1890s (*Pop. Trends* 1990). Fieldwork for the BTO *Atlas* (1976) recorded breeding only along the Severn valley west of Ironbridge, along the Teme valley, in the Oswestry uplands and in the hill country to the south-west. Including those on the borders, a total of 21 10-km squares were occupied then compared with 37 now. Many of the 10-km squares colonised since 1972 have several occupied tetrads, and breeding has also been noted at a number of sites in the lowland north, demonstrating a significant increase in range and numbers.

The provision of nestboxes, which can increase breeding density and encourage the colonisation of other woods, is partly responsible for this increase but the extent is difficult to assess. Pied Flycatchers take readily to nestboxes in woods that meet their strict habitat requirements, preferring them to naturally occurring holes. In the Clun valley 12 nestboxes were erected in Sowdley Wood in 1979 and by 1989 over 1800 nestboxes had been provided along the river and its tributaries, and in adjoining woodland, with up to 30% occupied by Pied Flycatchers in a good year. A further 710 boxes have been put up in the valleys around The Long Mynd and there are more at other sites in the south. The total number of nestboxes in known schemes in south Shropshire currently exceeds 2700 but further expansion is not envisaged.

Ringing has indicated that recruitment to this area has chiefly been from Hereford and Worcester, and Powys in Wales. Numbers vary from year to year, reflecting losses on migration and in the winter quarters as well as breeding success. Ringing recoveries of nestlings or locally breeding adults have occurred in France, Spain and Morocco, the majority reported as "killed by man". Unseasonally late frosts or a poor caterpillar crop can have a disastrous effect on breeding success, as in 1990 when almost 25% of the nestlings died of starvation in the nestboxes before fledging. Ringing has also shown that Pied Flycatchers, particularly females, are not faithful to their natal area and have been caught at nestboxes the following year in Yorkshire, South Wales and the West Country, whilst others have appeared in the breeding season in Belgium, Holland and Norway. There is some evidence to suggest that in years when numbers of Blue and Great Tits are high, Pied Flycatchers retrapped in south-west Shropshire originated from further south in England or Wales, whilst in years when tit numbers are low more retraps are from northern England (J.M. Langford and C.J. Whittles, pers. comm.)

Nationally the population level is believed to be stable, with nestbox schemes confusing the picture in some areas (*Pop. Trends* 1990). Some of the local schemes mentioned above had high occupancy levels in 1990 following good breeding success in 1988 and 1989. In one scheme in 1989, 413 nests produced 2128 fledged young. Some tetrads may naturally hold more than 20 pairs and, where nestboxes are liberally provided in suitable habitat, up to 50 pairs may occur in good years. Away from these strongholds the numbers will be much lower, and taking an average of 10 pairs per occupied tetrad gives a population of about 2000 pairs, though it may be higher in good years.   *CEW*

LONG-TAILED TIT
Aegithalos caudatus

Status: **Resident**

Tetrads with evidence of breeding

Confirmed	478 (55%)
Probable	147 (17%)
Possible	102 (12%)
Total	**727 (84%)**

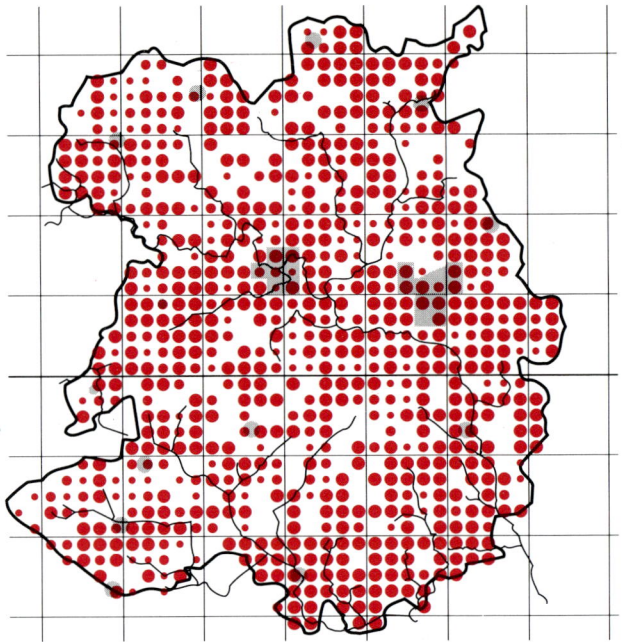

The Long-tailed Tit's nest is one of the wonders of bird architecture, being both beautifully designed and skilfully constructed. Built in the fork of a tree, in thorny bushes or in brambles, the result of its predation by Jays or Magpies is one of the saddest sights in birdwatching. Breeding occurs on the outskirts of woods, in woodland clearings, in avenues of trees along streams, and sometimes in hedgerows. It is rarely found in urban gardens, or open land with poor hedges or scattered trees. The gaps therefore mainly correspond to towns, higher ground and arable farmland, although it could have been overlooked in tetrads only visited after hard winters in the first few years of the survey, when they may have contained much-diminished populations.

The best evidence of territory is found early in the season when nest-building by the pair is under way, and breeding is confirmed when both parents take food to the nest or when family parties suddenly become obvious, usually about the end of May. As eventually these parties may move considerable distances, sightings later in the summer do not prove breeding nearby.

Several major factors affect the population. Most importantly, numbers depend heavily on preceding winter weather conditions and extended periods of hoar-frost may even wipe out the population in some areas. Also, each territory holds an adult pair plus other adult helpers, so estimates are of territories rather than pairs. Finally its inconspicuous behaviour is likely to lead to under-estimation, particularly in woodland. Working from an estimate of 200,000 territories in Britain, perhaps an average of around 100 per occupied 10-km square, and a regional density considered to be around or rather lower than the average (*Pop. Trends* 1990), there are probably no more than around 2500–3700 territories, equivalent to 4–6 per occupied tetrad, even at the relatively high level at the end of the Atlas period.   DS

MARSH TIT
Parus palustris

Status: **Resident**

Tetrads with evidence of breeding

Confirmed	159 (18%)
Probable	131 (15%)
Possible	119 (14%)
Total	**409 (47%)**

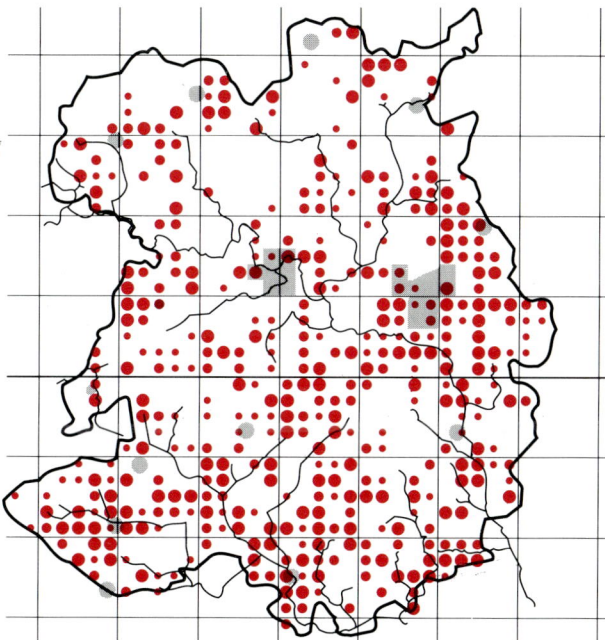

Marsh and Willow Tits are present throughout the year. They are easily separated from other members of the tit family but distinguishing between them is very difficult. Subtle differences in structure and markings are not always apparent, especially when the birds are moving or glimpsed briefly. The "pitchou" note of the Marsh Tit and the buzzing call of the Willow Tit are distinctive and often indicate presence before they are seen. When silent it is not always possible to make a positive identification. Records received for "Marsh/Willow Tit" were discounted, but it is likely that some errors in identification were also made.

Breeding may be proved for both species when young are being fed, or when the family party has left the nest hole. Both are thinly distributed and elusive as shown by the large proportion of possible and probable breeding records. The Marsh Tit reaches its highest densities in deciduous woodland, but also occurs in copses and hedgerows, and is widely distributed. It was found in every 10-km square, which shows a slight improvement since the BTO *Atlas* (1976), and the number of *SBR* records has also more than doubled over the last five years while the number of recorders has remained constant. This may indicate an increase in the Marsh Tit population or it could be due to the more detailed nature and extra fieldwork generated by this Atlas. Nationally there has been a shallow decline in numbers since the late 1960s (*Pop. Trends* 1990). If the estimated national average of 50–100 pairs/10-km square (BTO *Atlas* 1976) has not altered significantly and applies to Shropshire the population is probably between 1750 and 3500 pairs.   *APD*

WILLOW TIT

Parus montanus

Status: **Resident**

Tetrads with evidence of breeding

Confirmed	104 (12%)
Probable	119 (14%)
Possible	114 (13%)
Total	**337 (39%)**

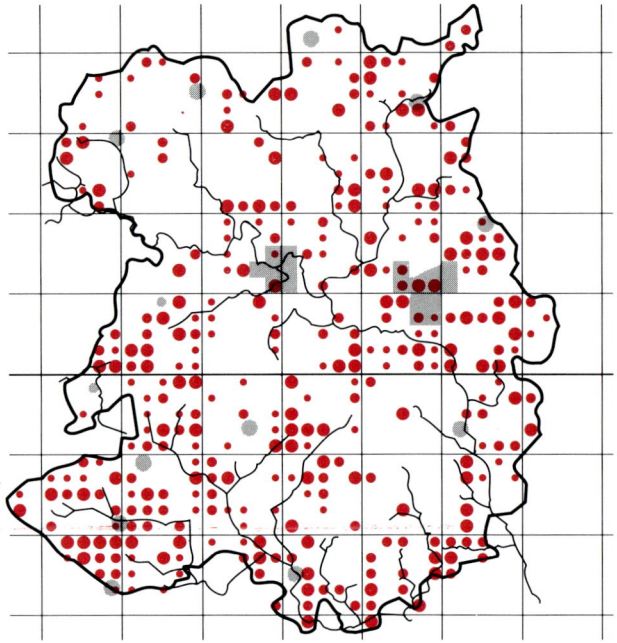

The Willow Tit was apparently not recorded in Shropshire until 1945 (*Handlist* 1964), almost fifty years after it was first recognised as a British bird. Previously it had been overlooked due to its similarity with the Marsh Tit (for differences see Marsh Tit).

Usually found in damp woodland or carr, the Willow Tit is the only member of the tit family found locally that excavates its own nest hole. Suitable rotten timber may be more readily available along river valleys and many records came from such tetrads (see Map 2). It is less dependent on deciduous woodland than the Marsh Tit, taking readily to conifer plantations, and often occurring in scrub and farmland where quite small damp areas exist.

The Willow Tit is generally the scarcer of the two species, producing less *SBR* records and found in fewer tetrads. The south-west is an exception, and in the final year concentrated Atlas fieldwork in this area produced a majority of Willow Tit records for the first time (*SBR* 1990). This may be due to the large number of conifer plantations, higher rainfall or the amount of rotten timber left standing in this difficult and isolated terrain. Further research on the population and habitat of the Willow Tit in this area would be welcome.

It was present in every 10-km square, which shows an increase since the BTO *Atlas* (1976), but this may be due to more detailed coverage. Nationally numbers have remained stable except for minor fluctuations related to winter weather conditions (*Pop. Trends* 1990). Assuming the estimated national average of 40–80 pairs/10-km square applies, the population is probably between 1400 and 2800 pairs. *APD*

COAL TIT

Parus ater

Status: **Resident**

Tetrads with evidence of breeding

Confirmed	344 (40%)
Probable	147 (17%)
Possible	148 (17%)
Total	**639 (73%)**

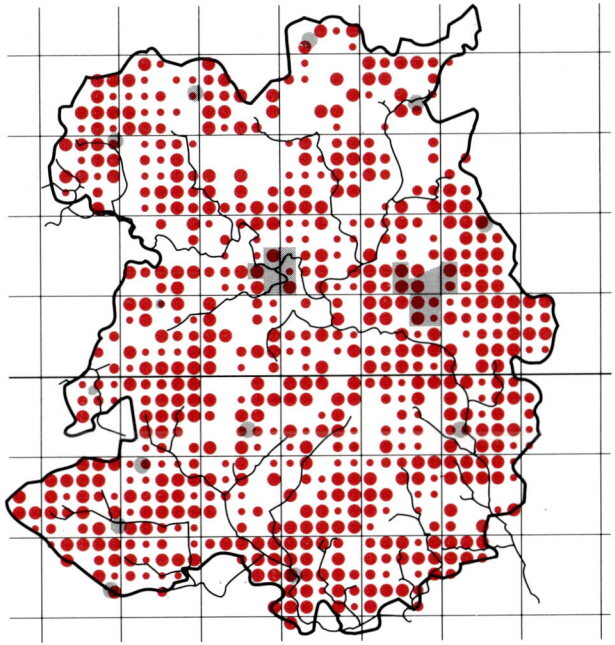

The highly sedentary Coal Tit particularly favours coniferous woodland, but is also found at lower densities in mixed and deciduous woods; even relatively small groups of conifers are sufficient. Nests are made in any convenient hole in a tree, bank or wall, including nestboxes. It tends to be absent from hedgerows, scrub and scattered trees which are attractive to Blue Tits, and relative shyness ensures smaller gardens are avoided. These habitat preferences are reflected in an absence of Coal Tit records from cereal-growing and moorland areas.

Presence is usually detected from calling but, as the nest is often well hidden, evidence for confirmed breeding usually comes from adults feeding young. Whilst difficult to overlook in prime habitat because of territorial competition, it may well go unrecorded in areas with only the odd pair.

Based on national CBC figures equivalent to around 200 pairs per 10-km square for farmland and woodland (*Pop. Trends* 1990), the population would be somewhat lower than 8000 pairs. However, the increased planting of conifers over recent times has undoubtedly improved its breeding status and densities in favoured woods may be as high as 100 pairs per sq. km. As around 4% of Shropshire is now coniferous woodland, the population may be as high as 20,000 pairs. *DS*

BLUE TIT
Parus caeruleus

Status: **Resident**

Tetrads with evidence of breeding

Confirmed	843 (	97%)
Probable	19 (	2%)
Possible	7 (	1%)
Total	**869 (**	**100%)**

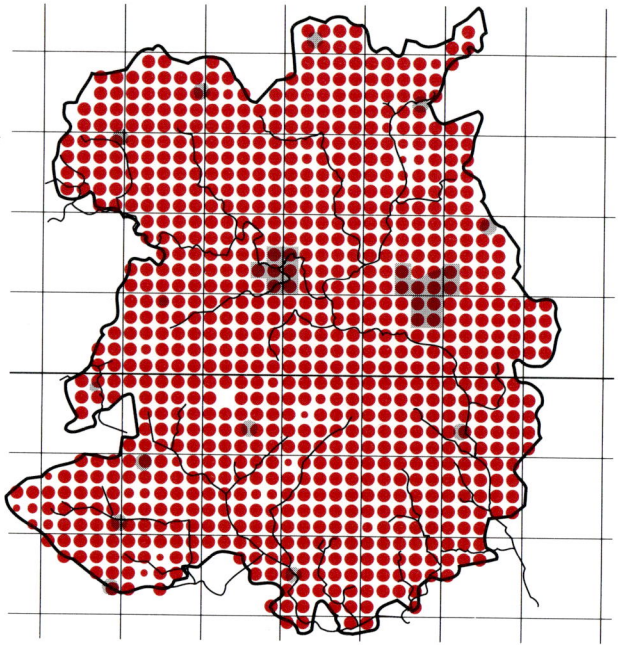

JS

Blue Tits are much more common than the other members of the tit family. Regular visitors to gardens, their acrobatic feeding behaviour generates considerable popular interest in birds.

Though primarily inhabitants of deciduous woodland, they breed wherever there are suitable nest holes. Trees provide the vast majority, but other holes — in large bushes, buildings and nestboxes — are also used. Even intensively farmed and urban areas contain suitable nest sites, though breeding success may be reduced there. The only relatively large areas devoid of suitable habitat are the open heather moorlands. As the hillsides and stream valleys have scattered trees, small woods and hawthorn hedges, there are no tetrads without some suitable breeding habitat.

Breeding is easy to confirm. Caterpillars and other larvae are clearly visible in the beaks of both adults as they continuously convey food back to the nest for the large brood, usually of 7 to 10, for 17–18 days. Fledglings continue to be fed for some days after they leave the nest, and family parties are kept together by repeated call notes, making them easy to locate and identify.

The British population is estimated at 3.5 million pairs, an average of roughly 15 pairs/sq. km, but western England has a relatively high density, and though Great Tits generally number about 60% of Blue Tits (*Pop. Trends* 1990), experience of Atlas fieldwork suggests the ratio of Blue Tits is generally higher here, and the population is likely to be at least 60,000 pairs. *LS*

GREAT TIT
Parus major

Status: **Resident**

Tetrads with evidence of breeding

Confirmed	759	(87%)
Probable	60	(7%)
Possible	34	(4%)
Total	**853**	**(98%)**

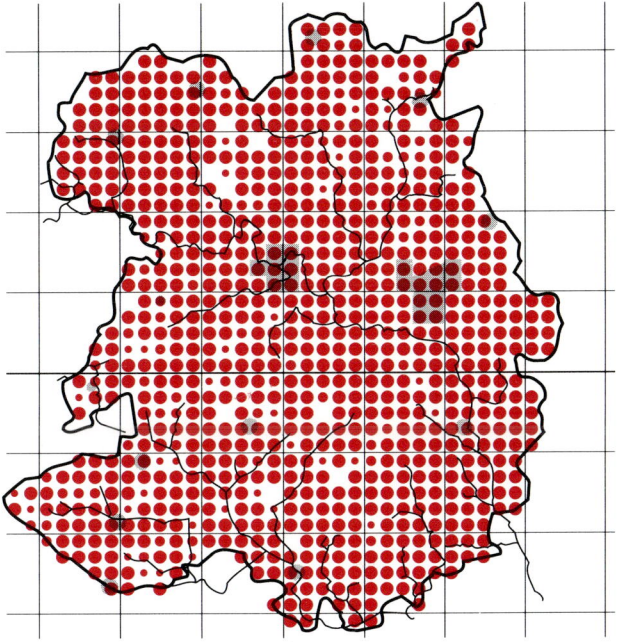

The Great Tit is a well known and common resident, favouring deciduous woodland, but tending to avoid areas of farmland and upland with few trees, where there is relatively little suitable habitat. in tetrads which consist largely of prairie-style arable farming it is very scarce, and though it may have been overlooked it may even be absent from a few of them altogether.

Distribution and behaviour are similar to the Blue Tit's and breeding is confirmed in the same manner. However, the Great Tit is not so agile, requires a larger, less frequently occurring nest-hole and a larger territory, and fluctuations in population caused by hard winters are more pronounced. Consequently it is less numerous in the Blue Tit's more marginal habitats and its national population levels are generally about 60% of those of its close relative (*Pop. Trends* 1990).

Greater provision of nestboxes has undoubtedly improved its breeding range and density, for example in the Clun Forest, and post-boxes are taken over in several places, requiring concerned owners to make alternative arrangements for their mail.

As the national population is estimated at 2 million pairs, averaging roughly 8 pairs per sq. km, the corresponding figure for Shropshire is around 30,000 pairs.   DS

NUTHATCH
Sitta europaea

Status: **Resident**

Tetrads with evidence of breeding

Confirmed	283 (33%)
Probable	169 (19%)
Possible	171 (20%)
Total	**623 (72%)**

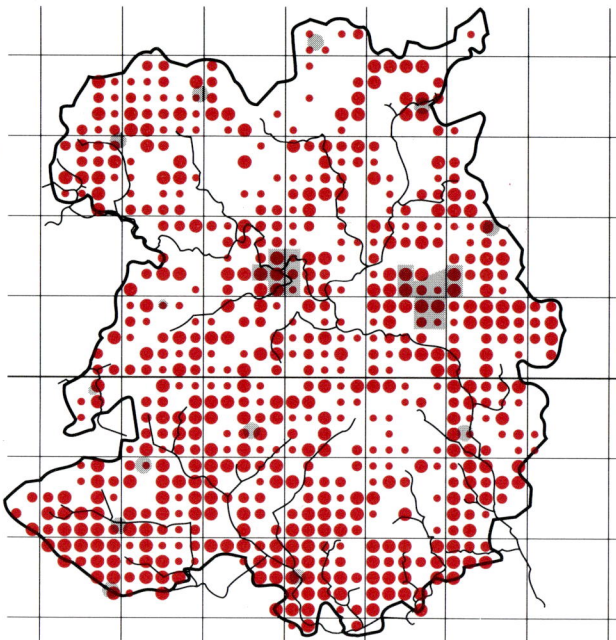

The sleek Nuthatch gives the appearance of a small woodpecker as it moves with short, jerky leaps on tree trunks and boughs.

Nuthatches nest in holes and the optimum habitat is mature deciduous woodland, especially woods containing oak, beech and hazel which provide an important food source. They are also found in other wooded areas with mature trees such as parkland, well-timbered hedgerows and small copses, and will use nestboxes in suitable habitat.

In early spring the loud metallic ringing calls and characteristic song make initial location easy, and holes in trees in the close vicinity can provide evidence of occupation as the Nuthatch is unique amongst British birds in that it reduces the size of the entrance using mud. Occupied nestboxes can also be identified by this behaviour: the entrance may be the optimum size but the urge is so great that mud is deposited around the lid or even just dumped on top.

Nuthatches are generally widespread, with the highest densities of breeding pairs in well-wooded areas. There are significant gaps in their distribution in the north, west and south-east, which correlate to a lack of suitable woodland in what are mainly agricultural areas (see Maps 7, 8 and 12). They are also absent from the open moorlands, but do extend into the uplands if there are suitable woods. Uninhabited tetrads in otherwise well-populated areas stand out as islands with unsuitable breeding habitat. The BTO *Atlas* (1976) showed two 10-km squares without Nuthatches and another seven with only possible breeding. The current map has confirmed breeding in every 10-km square, reflecting the apparent doubling of the British population since then, mainly because of the effects of Dutch elm disease (*Pop. Trends* 1990).

By virtue of their sedentary nature, possible and probable records almost certainly refer to breeding pairs. Based on impressions gained during Atlas fieldwork an average of 5–10 pairs per occupied tetrad seems reasonable, giving a population of around 2250–4500 pairs.

JM

TREECREEPER
Certhia familiaris

Status: **Resident**

Tetrads with evidence of breeding

Confirmed	313 (36%)
Probable	175 (20%)
Possible	230 (26%)
Total	**718 (83%)**

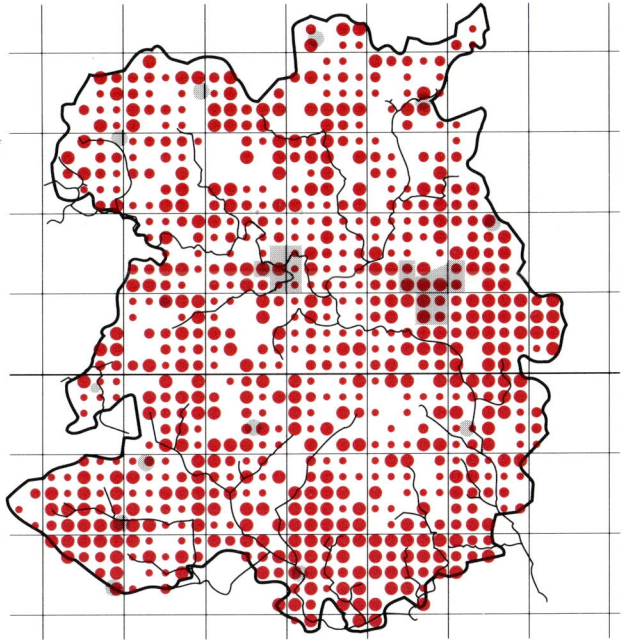

Small and well camouflaged, the Treecreeper is one of the more inconspicuous woodland birds. It often first catches the eye alighting on a trunk and then climbing, mouse-like, upwards or along branches, purposefully probing for insects and spiders within bark fissures; and then dropping down to the base of another tree to recommence its upward journey. Suitable wooden fences are also worked in similar fashion.

The Treecreeper can be found in all types of woodland but is most abundant in mature deciduous or mixed woods which provide nest sites behind loose bark or in the crack of a broken branch. These nesting requirements probably act as a limiting factor in some types of woodland, such as young conifer plantations, but it may occur almost anywhere with mature trees, such as hedgerows, tree-lined waterways, parkland and large gardens. Reflecting this diversity of habitat, the Treecreeper is widespread, but absent from those areas which have been cleared for intensive agriculture, and treeless areas of upland, though high altitude is no barrier in itself.

The song period starts in early spring but is of short duration, and the song itself is high pitched and may be difficult to hear amongst that of other birds. It is thus easily overlooked, and may have been missed in some tetrads which appear to have suitable habitat. Obtaining evidence of probable or confirmed breeding is difficult, but two broods are often raised, and adults feeding youngboth in and out of the nest provided a high proportion of the confirmed breeding records.

The breeding population decreases after hard winters if ice-glaze covers tree trunks and branches for long periods. Numbers were reduced after 1962–63 (*Handlist* 1964) but not apparently at the start of this Atlas period. The Treecreeper is sedentary, so possible and probable records will almost certainly refer to breeding pairs, though densities may generally be lower in these tetrads. Assuming there are 10–20 pairs per occupied tetrad, and allowing for a small degree of under-recording, the population is estimated at between 5000 and 10,000 pairs.   JM

JAY

Garrulus glandarius

Status: **Resident**

Tetrads with evidence of breeding

Confirmed	197 (23%)
Probable	287 (33%)
Possible	216 (25%)
Total	**700 (80%)**

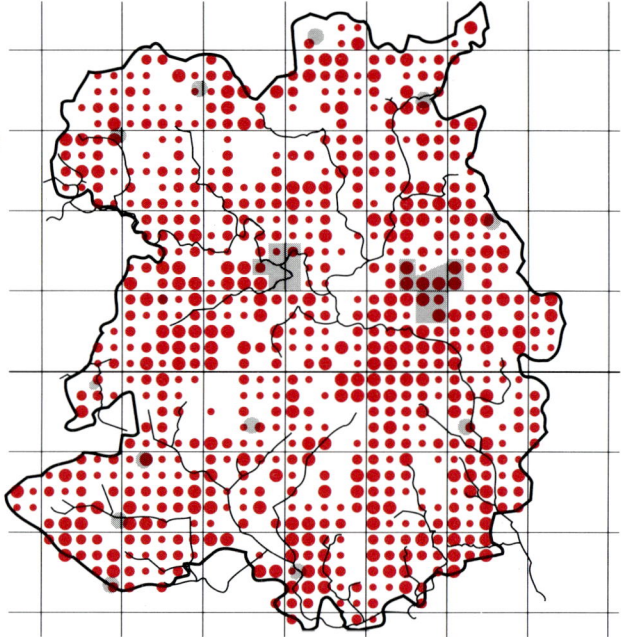

By far the most secretive corvid, and the one most restricted to woodland, the Jay is surprisingly difficult to locate despite its large size and striking colours. Oakwood, which it helps to regenerate by burying acorns for winter food, is the preferred habitat. Most other deciduous and coniferous woods are also used, although the latter only support lower breeding densities (see Map 8). Scattered trees and small copses are not usually suitable.

Nests are built in thick cover, and most food — large insects, fruit, nuts, seeds, and the eggs and young of other birds — is gathered within the wood. Much of the habitat is therefore difficult to survey and Jays do not defend a territory, so sightings usually consist only of an occasional glimpse. This secretive behaviour and a highly sedentary nature suggest some pairs have been overlooked, and most of the possible and probable records will in fact relate to nesting pairs. Breeding has been confirmed infrequently, usually by chancing across a family party. The large gaps in the distribution correspond to areas of open hillside and moorland, and heavily cultivated areas with little woodland.

In common with other corvids, persecution by gamekeepers kept numbers low until the First World War, after which they increased steadily, more recently as a result of commercial afforestation. Population levels do not seem to be affected by hard winters (*Pop. Trends* 1990).

The average breeding density found on CBC woodland plots in 1972 was 5.1 pairs per sq. km (BTO *Atlas* 1976), and this has increased by almost 25% since (*Pop. Trends* 1990). Assuming this applies to Shropshire woodland, which is 7.2% of the area, the population is estimated at around 1600 pairs, or just over 2 pairs per tetrad.   *LS*

MAGPIE
Pica pica

Status: **Resident**

Tetrads with evidence of breeding

Confirmed	767 (	88%)
Probable	79 (	9%)
Possible	20 (	2%)
Total	**866**	**(100%)**

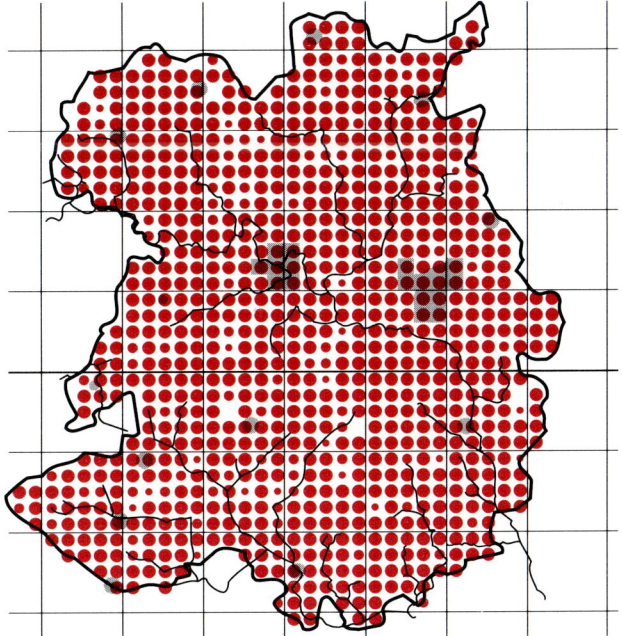

The aggressive cackle of the marauding Magpie characterises it as one of our least popular residents. Though not found in extensive woodland, it occurs wherever there are trees or bushes for the nest and open ground for feeding, from large gardens in towns, through all types of farmland to the upper valleys of the highest hills.

Breeding is easy to confirm. Magpies are very conspicuous, and noisy behaviour attracts attention to the obvious bulky nest, usually in isolated trees or hawthorn bushes in hedgerows. Except perhaps in deep cover, it has a distinctive domed roof, presumably designed for defence against the much larger Carrion Crow.

The Magpie "has greatly increased since 1940" (*Handlist* 1964), particularly by expanding into suburban areas and more open countryside. Before then it had been controlled by gamekeepers, especially before the First World War, as it eats the young and eggs of game and other birds. In fact it is omnivorous, eating mainly invertebrates and insects, but also small mammals, carrion and berries, so a wide range of habitats are occupied. It undoubtedly breeds in every tetrad. The population has continued to increase, and nationally CBC results suggest there are now around 5 pairs per sq. km (BTO *Atlas* 1976 estimate, doubled in line with the index in *Pop. Trends* 1990), or 20 pairs per tetrad, giving an estimate of around 15,000–20,000 pairs if applied locally.

Many people told Atlas fieldworkers they believed the increase in Magpies was responsible for a decrease in song and garden birds. Several had witnessed the devouring of eggs or nestlings, including one case where even a much larger incubating Pheasant was driven from her nest. Considerable research is now being done on the effect of growing Magpie numbers on other species. However, the evidence that already exists indicates that even in summer Magpies are essentially ground feeders, and no correlation has been found between the Magpie's increase and the breeding success of any small garden bird (*Pop. Trends* 1990). Other more insidious factors, such as habitat destruction, the use of insecticides and weedkillers by gardeners, and predation by domestic cats, have a devastating effect while the highly visible Magpie gets the blame.

LS

JACKDAW
Corvus monedula

Status: **Resident**

Tetrads with evidence of breeding

Confirmed	649 (75%)
Probable	113 (13%)
Possible	68 (8%)
Total	**830 (95%)**

JS

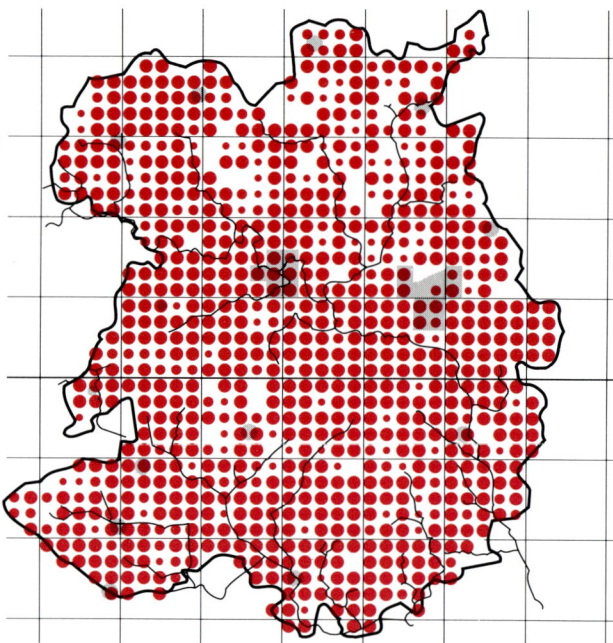

The distinctive call together with smaller size and ash-grey nape makes the Jackdaw easy to distinguish from the Rook and Carrion Crow. A common and widespread resident usually occurring in small colonies, it nests in holes in quarries, old buildings and large trees, and in chimneys in suburban areas. Breeding can be confirmed by watching adults going into the nest site.

A mixture of grassland and arable land is preferred and Jackdaws spend much time feeding on grass, but they will regularly scavenge on rubbish tips and during the breeding season take caterpillars and other larvae from the leaf canopy. Where the colonies are few and far between or obscured by a rookery they could be overlooked and this may explain some of the gaps. Lack of suitable nesting and feeding areas may also be a factor as many of the gaps and small dots coincide with intensely arable areas, moorland and urban Telford.

The BTO *Atlas* (1976) gave an estimate of about 6 pairs per tetrad but, like the Carrion Crow, the national population has probably doubled since that time (*Pop. Trends* 1990). Taking an average of 12 pairs per occupied tetrad suggests a population of almost 10,000 pairs.   *CEW*

ROOK
Corvus frugilegus

Status: **Resident**

Tetrads with evidence of breeding

Confirmed	545 (63%)
Probable	48 (6%)
Possible	142 (16%)
Total	**735 (84%)**

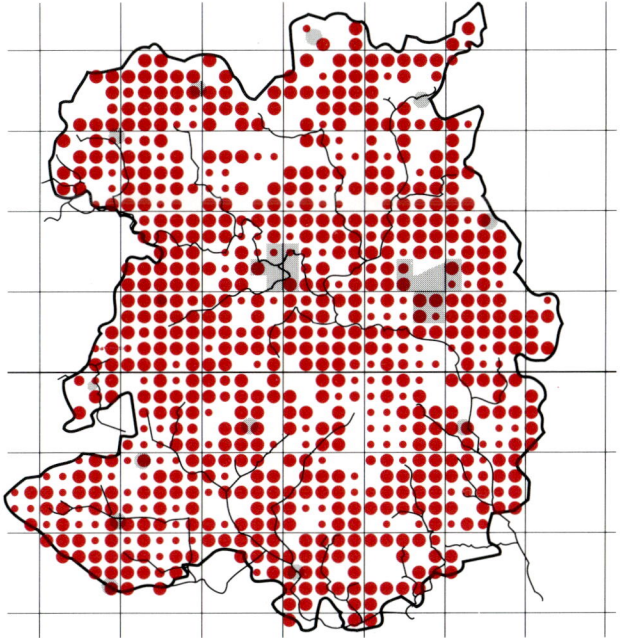

The sound of a busy rookery early in the spring is, aesthetically, an essential ingredient of the English countryside. Rookeries are quite easy to locate at that time and nearly three-quarters of the records are of confirmed breeding. Once the leaves are open or where nests are in conifers, small sites may be overlooked. Foraging may take place some distance from the colony, especially by family parties after the young fledge, so some of the smaller dots may relate to breeding in adjoining tetrads.

Widely distributed, the Rook favours a mixed grassland–arable regime. The increased area now under cereals combined with the mechanisation of harvesting has had an effect on Rook ecology by changing the times of year in arable areas when food will be abundant (O'Connor & Shrubb 1986). Although the resulting decline is more evident in eastern England these changes may explain some of the gaps in north Shropshire (see Map 7). Other gaps correlate with the towns, large woods and higher uplands with few trees although there are some rookeries above 300m, the highest being on the Stiperstones at 350m (*SBR* 1987). Most have been occupied every year of the Atlas period as Rooks are very faithful to traditional sites. It normally takes tree felling to make them move elsewhere, but trees weakened by disease, age or gale damage will be abandoned, and regular shooting has eliminated at least one rookery.

A national BTO census in 1975 located 460 rookeries containing 12,092 nests in the county (Sage & Vernon 1978). These rookeries were contained within 292 tetrads, giving an average of 41.4 nests per tetrad. A sample census in 1980 showed a slight decline in the number of rookeries but an increase in their size, resulting in an increase in the total number of nests (unpublished BTO data for Shropshire). Now, with confirmed breeding in 545 tetrads and assuming an estimated 45 pairs per tetrad, the breeding population will be about 25,000 pairs. The 1975 census revealed a serious decline in Rook numbers nationally over the previous twenty years but this has now halted (*Pop. Trends* 1990). Our figures suggest a modest recovery in Shropshire but care should be taken when comparing counts from a single season census with the results from a 6-year study which will have covered the ground more thoroughly.

CEW

CARRION CROW
Corvus corone

Status: **Resident**

Tetrads with evidence of breeding

Confirmed	767 (	88%)
Probable	73 (	8%)
Possible	26 (	3%)
Total	**866 (**	**100%)**

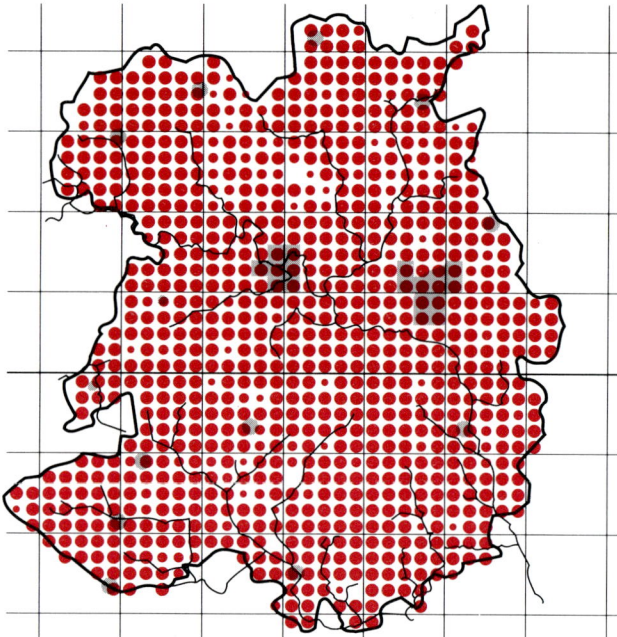

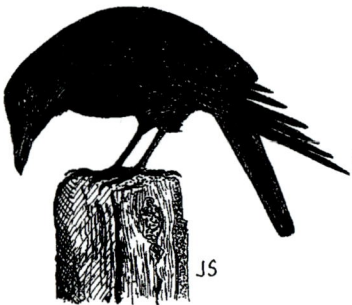

The Carrion Crow is one of the most widespread species, recorded in all habitats provided there is at least one substantial tree, the most usual site for its nest. In rural areas hedgerow trees and shelterbelts are used, whilst in suburban areas specimen trees in parks and gardens are popular. In the uplands a large hawthorn will suffice if nothing larger is available and the conifer plantations of the south provide further sites.

Crows start breeding early before the leaves emerge, so nests are easily found and most of the records are of confirmed breeding. Leaf cover makes nests difficult to locate and absence of confirmed breeding may relate to areas which were only visited later in the season. Crows are unpopular with sheep farmers and gamekeepers and in some areas shooting and illegal use of poisons may reduce the population, making them more wary and difficult to find. Generally they are tolerated, especially in the arable areas and around the towns.

Being so common there are few *SBR* reports, which usually refer to large flocks or unusually marked individuals. Most are winter flocks but, in May 1985, 600 fed together in one meadow near Westbury, close to the Long Mountain.

Though still persecuted, the population has grown considerably since the First World War and more rapidly since around 1940, following reductions in gamekeeping and more tolerance from farmers. The *Handlist* (1964) reported "numbers have increased considerably in recent years". The BTO *Atlas* (1976) gave an average of 10 pairs per tetrad but since that time the national population has possibly doubled (*Pop. Trends* 1990). 20–25 pairs per tetrad would give a breeding population of between 17,500 and 22,000 pairs.   *CEW*

RAVEN
Corvus corax

Status: **Resident**

Tetrads with evidence of breeding

Confirmed	27 (3%)
Probable	72 (8%)
Possible	57 (7%)
Total	**156 (18%)**

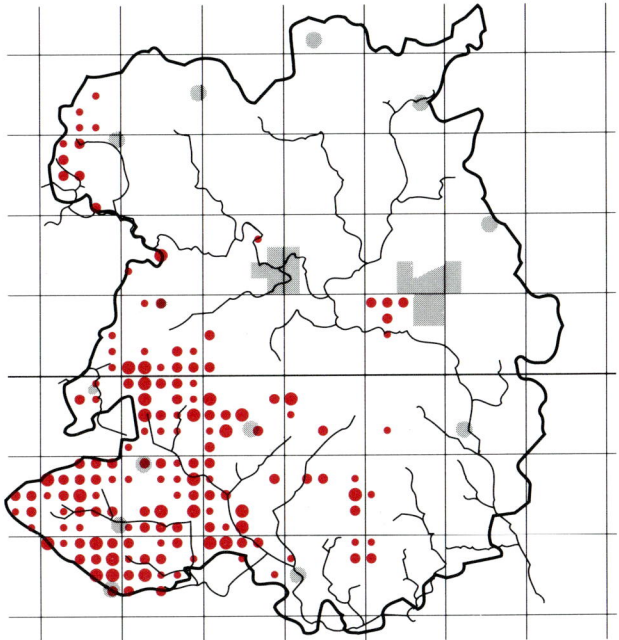

Ravens probably once nested throughout Shropshire but 19th century persecution reduced the population to one pair, which last nested in 1884 in a disused quarry site at Ashes Hollow on The Long Mynd. In 1918 a pair returned to the "identical spot" (Forrest 1918) and a revival of Raven fortunes was under way.

The *Handlist* (1964) recounted that breeding was sporadic between 1918 and 1939, but that thereafter numbers increased steadily. By 1964 "about 20 pairs" were breeding regularly in the Clun Forest upland and west-central hills; The Wrekin and Clee hills were thought to be occupied too, while they were "regularly seen" in the Oswestry upland where nesting had occurred in the 1940s. Other locations mentioned in *SBR*s are Millichope, Ashford Carbonell and Bridgnorth. The map in the BTO *Atlas* (1976) encompassed all these locations except The Wrekin and Bridgnorth.

Today's distribution shows only a few peripheral changes. There is a group of records in what may be a new area to the east of Clungunford, and confirmed breeding at Loton Park, a new site; Bridgnorth and Ashford Carbonell fail to figure but the Oswestry upland, The Wrekin, Clee hills and Millichope areas are all represented and the core area remains the south-west. Ravens principally inhabit the uplands where sheep farming is widespread on poor ground (see Maps 3–7 and 9). For nesting they favour conifers, especially Scots pines, and no cliff site is now used.

Breeding activity starts early in the year before Atlas fieldwork gets under way, but the nesting period is protracted and few pairs of this large and vocal species will have gone unrecorded, even if some active nests have been overlooked. However, equating each record with one pair would be an error, as territories may average in excess of 3 tetrads in size (Dare 1986), each territory will generally hold more than one nest site (Allin 1968), birds do not breed until they are 2 years old (Coombs 1978) and adult pairs may hold territory without breeding (Dare 1986). With these factors in mind, 30–35 breeding pairs is a reasonable population estimate.

Shropshire probably holds in excess of half the Ravens nesting in the English part of the Marches, and their fortunes ought to be followed closely, as what happens on this, the periphery of the Raven's range, may be indicative of what is to follow in Wales. Future trends may be downward, as improved animal husbandry is progressively reducing the sheep carrion upon which Ravens depend. *TW*

STARLING
Sturnus vulgaris

Status: **Resident**

Tetrads with evidence of breeding

Confirmed	818 (94%)
Probable	20 (2%)
Possible	22 (3%)
Total	**860 (99%)**

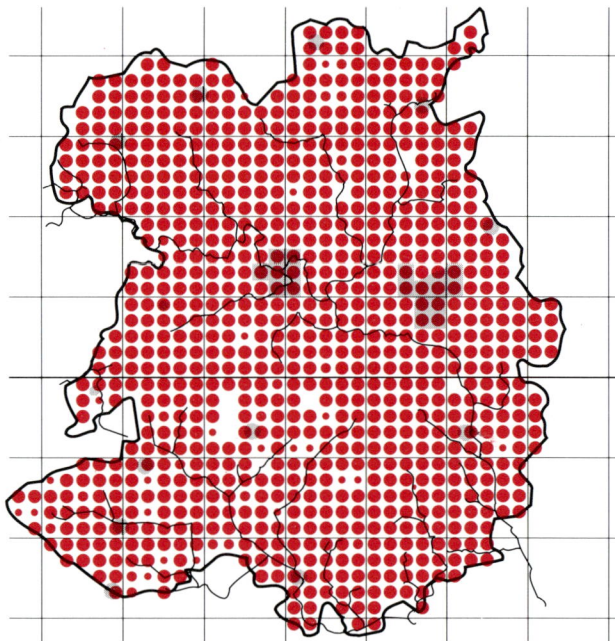

The abundant Starling has an extensive repertoire of calls and is an accomplished mimic. It breeds in urban and rural districts, on cultivated and uncultivated land and in deciduous woodland.

Starlings breed colonially or singly. Nests are built in almost any cavity: in chimneys, under eaves, in buildings, cracks, quarries, trees, haystacks, dense ivy, evergreens and even banks of streams. Occasionally-successful attempts to evict woodpeckers may occur after the excavation is completed. From mid-April the male commences nest construction which the hen completes. Breeding is readily confirmed by watching parents carry food to progeny or entering the nest site, the entrance of which is characteristically whitewashed with droppings. Many unpaired males continue to occupy potential nest sites during the breeding season, so care must be exercised to establish status. Usually one brood is reared, infrequently two.

The Starling is widely distributed, but breeding densities are low on exposed open moorland and intensively farmed arable areas, and it appears to be absent from the remote parts of The Long Mynd.

During the past 20 years the national Starling population has declined, due partly to a change in climate, and also to changes in farming practice. Greater use of insecticides and fungicides, re-seeding of pastures, and an increase of autumn-sown cereals with a consequent reduction in spring cultivation, have all diminished the populations of soil invertebrates, including leather-jackets, a much favoured prey (*Pop. Trends* 1990). The agricultural practices in Shropshire have followed this trend. Nationally the CBC population index for farmland is now around three-quarters of the figure of 1000 to 2000 pairs per 10-km square estimated in the BTO *Atlas* (1976), and if applied to the county suggests a population of 27,000 to 54,000 pairs.   MW

HOUSE SPARROW
Passer domesticus

Status: **Resident**

Tetrads with evidence of breeding

Confirmed	813 (93%)
Probable	32 (4%)
Possible	9 (1%)
Total	**854 (98%)**

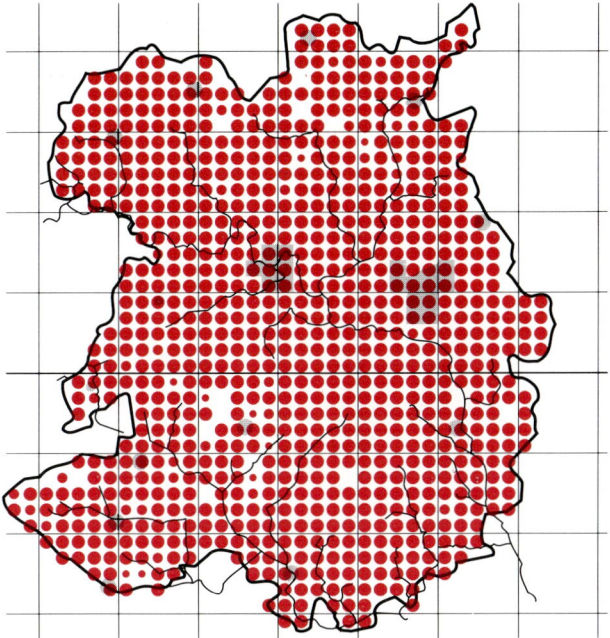

The *Handlist* (1964) recorded the House Sparrow as "resident and common throughout the county" but this statement has not been confirmed by the present study.

Closely associated with man, there is little difficulty in locating House Sparrows, and watching adults taking food into the nest hole, often under the eaves of a building, soon confirms breeding. Several broods are raised through an extended season and a nest with young was recorded as early as 12 January 1988 after a mild spell (*SBR*). They rely on man for much of their food, taking animal feed and spilt grain around farms and attending regularly at the suburban bird table. Where these supplies are not available they may be absent especially in the remoter areas. Nearly all tetrads contain human dwellings and only The Long Mynd is large and remote enough to contain tetrads where no people live, and House Sparrows are apparently absent. Elsewhere in the south and west much of the hill country is sparsely populated and House Sparrows are undoubtedly scarce, but sufficient habitat exists to hold a few pairs.

The BTO *Atlas* (1976) estimated 10–20 pairs per sq. km or 40–80 pairs per tetrad, which would give a population of not more than 70,000 pairs.

Pop. Trends (1990) indicates a possible decline in numbers especially in rural areas. One of the most common species in Britain, any change in population will need to be substantial before it will be noted in the field unless specific studies are undertaken.   *CEW*

TREE SPARROW
Passer montanus

Status: **Resident**

Tetrads with evidence of breeding

Confirmed	216 (25%)
Probable	178 (20%)
Possible	114 (13%)
Total	**508 (58%)**

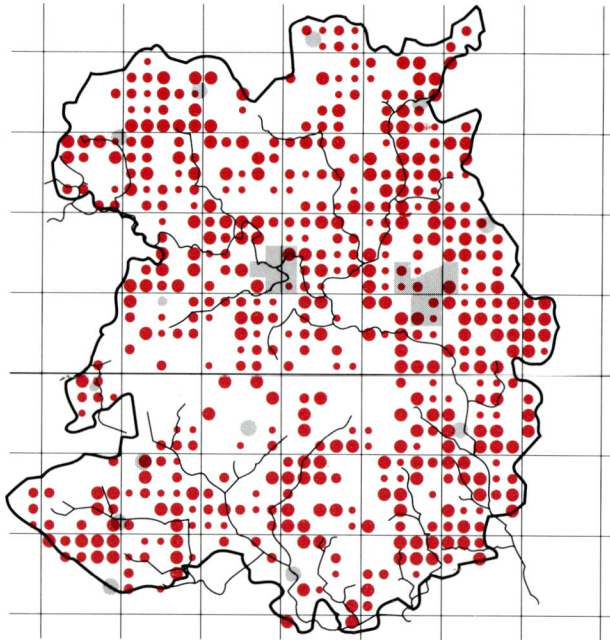

The *Handlist* (1964) recorded the Tree Sparrow as a "resident and non-breeding visitor, locally common as a breeding bird", and fieldwork for the BTO *Atlas* (1976) found Tree Sparrows in every 10-km square. The present study looks at the distribution in more detail and shows large areas where they appear to be absent.

Tree Sparrows are rare in upland areas but elsewhere are often found in loose colonies associated with groups of trees which provide nest holes. Nestboxes are used, and other sites include old nests of Sand and House Martins and isolated buildings where House Sparrows are absent. Breeding is best confirmed by observing adults entering nest holes or carrying food for young. As isolated pairs are secretive, the male and female appear identical, and recent fledglings look very similar to adults, establishing probable or confirmed breeding may be difficult, and all dots probably relate to breeding pairs.

Locating colonies is not easy even when the call is known and they may be overlooked, especially where mixed with House Sparrows. Some of the gaps may indicate low numbers rather than absence but although suitable breeding habitat appears to be available in most lowland areas gaps still exist, even where coverage has been good. One factor controlling numbers will be the amount of food available during the winter. Flocks that are reported then have usually been found on stubbles, a resource that is now rarely available for long before ploughing, and widespread use of herbicides has reduced the supply of weed seeds on farmland.

The national population is now strongly in decline though numbers increased dramatically in the British Isles from the mid-1950s to an estimated 850,000 pairs in the mid-1960s. Numbers then fell back, slowly at first but more rapidly in the late 1970s, to about 285,000 pairs in 1985. The earlier increases were probably a consequence of irruptions from continental Europe (*Pop. Trends* 1990). CBC data indicates that the decline has continued (Stroud & Glue 1991).

Atlas fieldwork suggests an average of 10 pairs in each tetrad where recorded, giving a population of about 5000 pairs.   *CEW*

CHAFFINCH
Fringilla coelebs

Status: **Resident**

Tetrads with evidence of breeding

Confirmed	747 (	86%)
Probable	109 (	13%)
Possible	12 (	1%)
Total	**868 (100%)**	

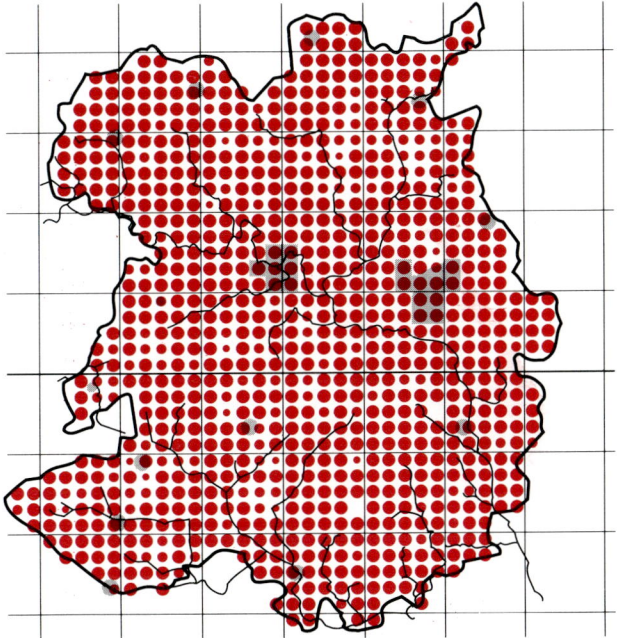

The loud distinctive song and colourful plumage make the male Chaffinch one of the most easily recognised harbingers of spring. Territorial behaviour starts in February, even though large numbers of this partial migrant may still be wintering here.

Chaffinches are very common and widespread. Altitude is not a constraint, and the only places where they may be absent are open areas with few hedges or trees to provide nest sites and song posts, but every tetrad contains suitable habitat. The delightful lichen-covered nest is ideally camouflaged, making it very difficult to see, and confirmation usually comes from adults carrying food.

After reaching a British peak in 1950, they were widely reported to be in decline in the late 1950s, with reduced breeding success probably due to agricultural chemicals, especially seed-dressings (BTO *Atlas* 1976). After an initial increase in 1962–66, a longer steadier increase continued from the mid-1970s (*Pop. Trends* 1990). British Chaffinches are sedentary, and numbers are periodically affected by hard weather and seasons with poor breeding results. Harsh weather occurred in the early months of 1985 and 1986, and cold wet springs in 1985–88, probably adversely affecting the breeding success of this normally single-brooded species.

The latest national population estimate is about 5 million pairs (*Pop. Trends* 1990) which gives an average density of less than 20 pairs per sq. km. Shropshire provides much better than the average national habitat and a figure of 40 pairs per sq. km would give a population of around 140,000 pairs. *JS*

GREENFINCH
Carduelis chloris

Status: **Resident**

Tetrads with evidence of breeding

Confirmed	368 (42%)
Probable	290 (33%)
Possible	129 (15%)
Total	**787 (90%)**

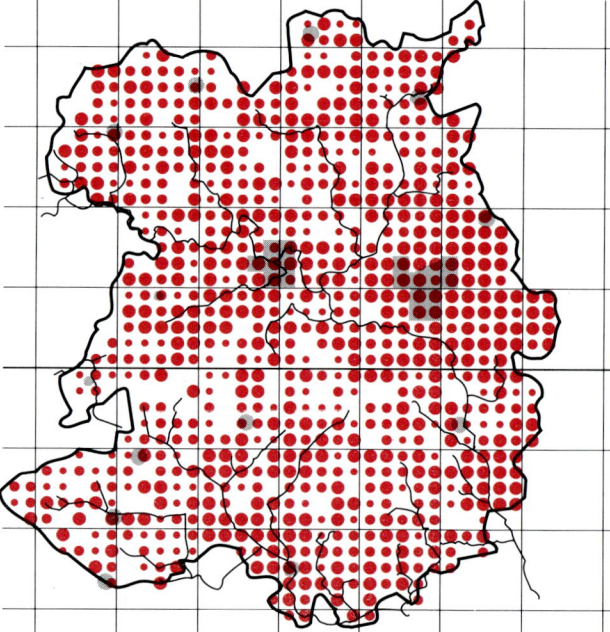

The distinctive "butterfly" display flight of the Greenfinch, and the long drawn-out nasal "zwee" of his song, attract attention to territories from early spring.

They are widely distributed, and may nest in loose colonies, but are nowhere numerous. More common at the turn of the century when Forrest (1899) reported them to be "abundant everywhere", the *Handlist* (1964) stated they were "found throughout the county" but "Considered by several observers to have decreased in the past twenty years". Greenfinches mainly inhabit the woodland edge, but in recent times have adapted to gardens and are especially fond of churchyards and vicarages, where they are attracted to yews. Open land at high altitudes, dense conifer plantations and large areas of improved pasture are unsuitable. Proving breeding is difficult as seed-seeking parents offer no visible evidence of carrying food to young, unlike insect-eating species, and this accounts for the low proportion of confirmed records.

Farmland numbers were at a low ebb in the early 1960s, probably due to organochlorine pesticides and the two severe winters of 1961–62 and 1962–63, but they had recovered by 1966. Woodland numbers have fluctuated since 1962, reaching a high point in 1986. With an estimated 800,000 pairs in Britain (*Pop. Trends* 1990) giving an average of roughly three pairs per sq. km, the population should be around 9000–10,000 pairs.

Future trends will be affected by farming practices. Greenfinches include much grain in their diet, and increased cereal production favours them, whilst the growing of rape, which they exploit, bridges an interval between early summer and the grain harvest. However, an increase in autumn sowing means fewer stubble fields through winter, making survival more difficult, and use of modern herbicides eliminates weed seeds, which are another important winter food (*Pop. Trends* 1990). These trends do not bode well for Greenfinches, though the recent vast growth in the number of households presenting garden bird food may have a significant effect on winter survival.   JS

GOLDFINCH
Carduelis carduelis

Status: **Resident**

Tetrads with evidence of breeding

Confirmed	340 (39%)
Probable	344 (40%)
Possible	97 (11%)
Total	**781 (90%)**

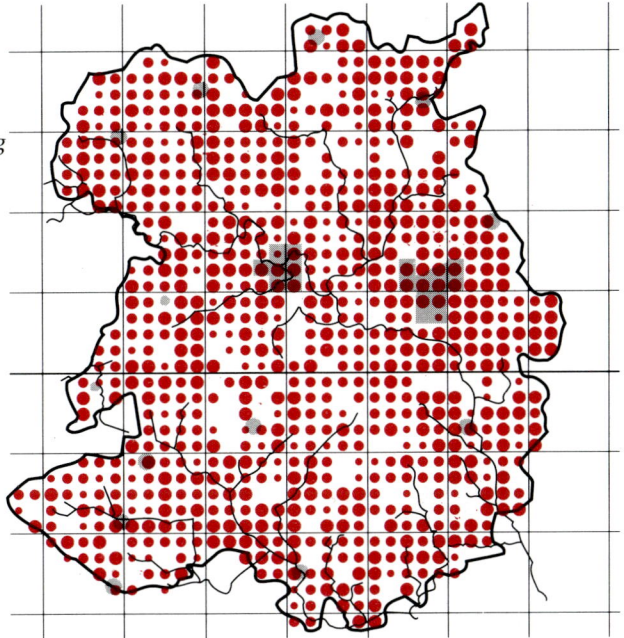

With both sexes in similar brilliant plumage, the dainty Goldfinch is a delight to the eye, and the collective noun "charm" is perfectly justified.

They breed in farmland, parkland, orchards, gardens, woodland edge and especially scrub and wasteland where there are trees and tall bushes to site the deep-cupped nest. They are absent from open ground at higher altitudes and areas lacking suitable nest sites. In spite of a protracted breeding season which allows up to three broods, breeding is difficult to prove as food cannot be seen in the bill. Sighting recently fledged young is the easiest way to obtain confirmation.

The national population was low during the 19th century, due to trapping Goldfinches for cage birds. A ban in 1881 brought a rise in numbers, but illegal trapping is still reported in the Shropshire press in present times, although the few now taken do not significantly affect the population. The *Handlist* (1964) reported them widely distributed and having "increased considerably during the last twenty to thirty years". The growth continued until the late 1970s. Losses from 1980 were probably due to herbicides and other agricultural changes affecting food supplies, but Shropshire appears to have suffered less than the south and east of England. There has been a partial recovery from the low point in 1986 (*Pop. Trends* 1990).

Territories are small, and pairs may breed in loose colonies, but nowhere are they numerous. Some distance may be travelled to gather food, which may have resulted in over-recording. With an estimated 300,000 pairs in Britain (*Pop. Trends* 1990), an average of somewhat less than 1.5 pairs per occupied sq. km, the population is perhaps up to 4500 pairs.

Although more efficient farming and a general tidying-up of wasteland has affected Goldfinches, their liking for tree seeds, especially birch and alder, and winter migration to France and Iberia by up to 80% of the population, should ensure that they are less vulnerable to agricultural change than other seed-eating species. Also the most favoured food, thistle, is less amenable to herbicide control than most other weeds.   *JS*

SISKIN

Carduelis spinus

Status: **Regular**

Tetrads with evidence of breeding

Confirmed	5 (1%)
Probable	18 (2%)
Possible	33 (4%)
Total	**56 (6%)**

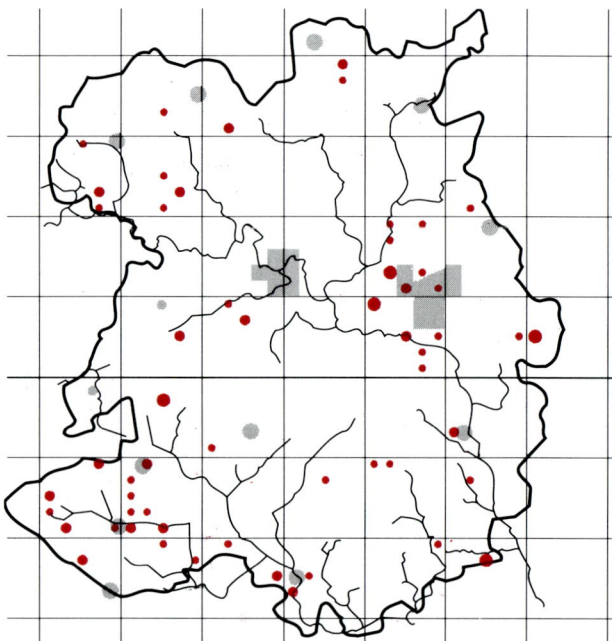

In recent years this delightful little finch has become a well-known garden visitor in late winter and early spring, when it appears to be particularly attracted to peanuts presented in red plastic bags. Some linger, such as a male singing regularly during the breeding season in 1984 in a Shrewsbury garden (*SBR*). Frequently sighted in April or even early May, many of the possible and probable records may be late migrants.

Breeding is always closely associated with conifers and they have adapted readily to plantations of introduced Sitka spruce. The widely scattered records, in the south and south-west from Ludlow to Newcastle, the north and north-west from Brown Moss to Llanymynech, and the east around Wellington, come from tetrads which contain either large plantations or scattered woods. Though breeding may be proved by hearing recently fledged young calling for food, the song is unfamiliar to most fieldworkers and the habitat is difficult to survey, so Siskins may have been overlooked, especially in extensively wooded areas such as the Wyre Forest and the larger plantations. Two of the confirmed records came from tetrads which contain only tiny patches of woodland (SJ61G and SJ80C).

"A pair nested at Grinshill in 1898, but the eggs were destroyed by a Jay" (Forrest 1899), and "Breeding was recorded in 1931, one pair and 1932, two pairs at Bucknell" (*Handlist* 1964). The BTO *Atlas* (1976) had no Shropshire records, and the next confirmed breeding was not until 1986 in the Wyre Forest (*SBR*). The general expansion into England from their traditional habitat in the Caledonian pine forests came after 1950 when the new conifer plantations were maturing to cone-bearing age. The range should continue to extend, helped by a combination of afforestation and a change in behaviour, taking advantage of garden feeding opportunities to aid winter survival (*Pop. Trends* 1990).

Siskins are eruptive, and the winter population varies widely. Migrants lingering beyond the irruption winter of 1985–86 (*SBRs*) may have given a false impression of high numbers in the following breeding season. However, in their main breeding areas elsewhere they often nest in loose colonies, so numbers here may be higher than the occasional sighting suggests. Colonisation is still at an early stage, and they are undoubtedly rare, so the population is unlikely to have reached 100 pairs.   JS

LINNET

Carduelis cannabina

Status: **Resident**

Tetrads with evidence of breeding

Confirmed	236 (27%)
Probable	334 (38%)
Possible	120 (14%)
Total	**690 (79%)**

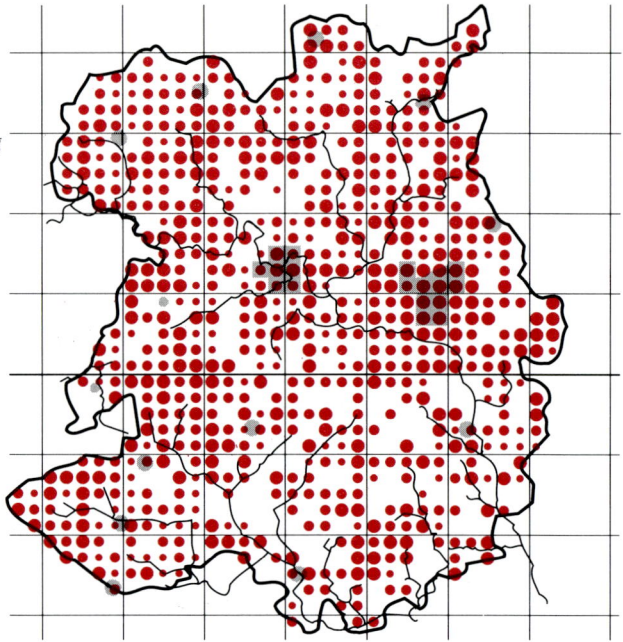

Linnets are sociable, and sizeable flocks are often seen in winter, though large numbers go to France and Iberia then, especially in hard weather.

They breed in small loose colonies in open landscapes, and pairs are easily located by the twittering flight call and constant display song. Nests are usually built close to the ground, in holly and hawthorn hedges in the lowlands and in gorse on upland and heathland, and may be easy to find, especially those built for first clutches in April and May. However, as Linnets are seed-eaters it is rare to observe adults carrying food to young, and fledglings can only be distinguished from the very similar females with great difficulty, so not surprisingly almost half the records remain at the probable breeding level.

Linnets are widely distributed, and the gaps generally correspond to heavily wooded areas, or those where suitable hedgerows or food plants are scarce.

They were a favourite cage-bird last century and trapping kept numbers low. Legislative protection in the 1880s allowed expansion, and agricultural recession between the wars provided an abundance of arable weeds, a major food source, so Linnets prospered until the post-war recovery increased the area of cultivated land again. The *Handlist* (1964) recorded that "Although found commonly throughout the County [the Linnet was] considered to have decreased in the past twenty years by several observers".

Nationally, following recovery to a peak after the hard winter of 1962–63, numbers have been falling since 1967, dramatically since 1977. This loss is due almost entirely to herbicides reducing the availability of weed seeds, resulting in lower breeding success and higher winter mortality. Increased weed control in Iberia, the winter quarters of part of the population, has also contributed to the steep decline here (*Pop. Trends* 1990).

Locally gorse, the favoured nesting bush in upland areas, suffered a massive die-back on The Long Mynd in the winter of 1981–82 (*The Flora* 1985). The shrub is only just beginning to recover in these higher locations, and this temporary loss of habitat may have affected numbers there.

The BTO *Atlas* (1976) estimated an average of 250–500 pairs/occupied 10-km square, but CBC data indicates a population decline of 70% on farmland since then (*Pop. Trends* 1990). Impressions gained during Atlas fieldwork suggest 3-6 pairs per tetrad may now be reasonable, giving 2000-4000 pairs in total.   *LS*

REDPOLL
Carduelis flammea

Status: **Resident**

Tetrads with evidence of breeding

Confirmed	7 (1%)
Probable	33 (4%)
Possible	48 (6%)
Total	**88 (10%)**

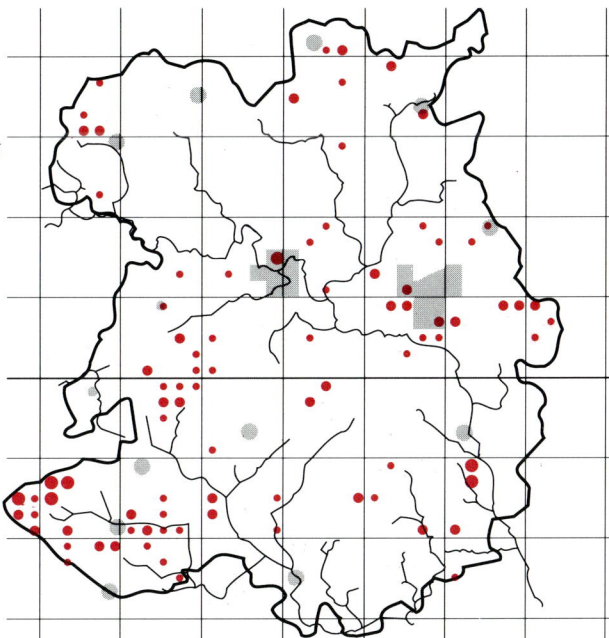

GH

The Redpoll is an elusive species for Atlas workers. It is mainly a partial migrant and winter visitor, often lingering late into May. Courtship display occurs before pairs move to their breeding grounds, mainly outside Shropshire, and these spring migrants account for most of the Atlas records.

By the time the breeding sites are reached, courtship has been largely completed so Redpolls are very unobtrusive there and easily overlooked. Pairs choose well-hidden sites, mainly in high hedges, especially old, thick hawthorns, and in coniferous woods.

The quiet contact call is the best means of locating them, but breeding is very difficult to confirm and has been proved in only seven tetrads, four of these being in the Clun Forest area, which does seem to have a small annual summer population (*SBRs*). The other confirmed breeding records, from Alkmond Park and Chelmarsh, appear to reflect isolated events at well-watched sites. Clusters of records in the Stiperstones area and around The Wrekin indicate the possibility of two other small pockets of breeding Redpolls and this is especially likely at the former, where a singing male was present on 16 June 1990 and two individuals in early July 1989.

The Redpoll seems never to have bred commonly, and although the *Handlist* (1964) recorded that nesting had been noted in all divisions, "reports of breeding have been few in recent years". Breeding was confirmed in a few more 10-km squares in the BTO *Atlas* (1976) than currently and Redpolls appear to have withdrawn from parts of the east since then.

Nationally the population is currently in decline (*Pop. Trends* 1990), after reaching what is thought to be the highest level this century in the late 1970s. Records are too scanty to determine whether a similar trend is true for Shropshire, but it seems that fewer than 100 pairs of this finch breed, and apart from the Clun Forest area, breeding appears to occur rarely at most sites (*SBRs*).   GT

COMMON CROSSBILL
Loxia curvirostra

Status: **Occasional**

Tetrads with evidence of breeding

Confirmed	6 (1%)
Probable	4 (1%)
Possible	15 (2%)
Total	**25 (3%)**

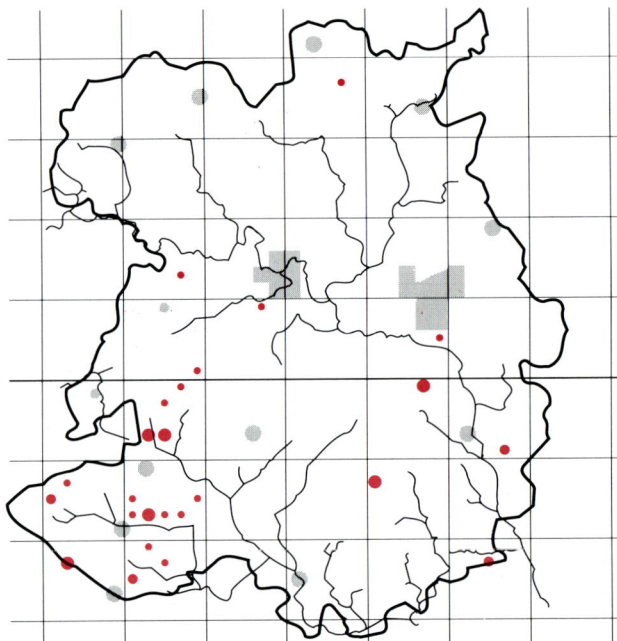

The irruptive behaviour of this fascinating bird clouds its status as a breeding species. It has been recorded annually in small numbers since 1976 and regularly before then, mainly as a winter visitor (*SBRs*).

The remarkable crossed mandibles are an adaptation so it can extract seeds from the cones of spruce, and to a lesser extent larch and pine. The Crossbill's dependence on these trees, which seed irregularly and locally, leads to occasional mass invasions ("irruptions") from northern Scandinavian and Russian conifer forests, as they seek cones with seeds. The proliferation throughout the UK of conifers, and spruce plantations in particular, has provided ideal habitat for these invading flocks and many now remain until the following spring.

Shropshire has a large acreage of spruces in the south-west, from where most Crossbill records originate, and breeding occurs in occasional years following irruptions. The difficulty of locating the nest, together with very early egg-laying, usually in February before fieldwork starts in earnest, has no doubt contributed to the dearth of breeding records.

Apparently only one Crossbill nest has been recorded (per W. Hotchkiss) and this was found during Atlas fieldwork near Shirlett in 1986. The five additional "proved" breeding records refer to the sighting of fledged young. Due to the unusually early breeding season family parties of Crossbills are on the move from June onwards, hence fledged young recorded at this time are unlikely to have bred locally. A female with a juvenile on 4 May 1985 at Bury Ditches, a much frequented site, is probably a genuine breeding record. So too is that from Brown Clee, where 3 pairs were seen together with 8 juveniles on 13 May 1986. Early spring records have come from there several times and a female and 3 juveniles were seen on 12 July 1988, so breeding may also occur on Brown Clee in other years. A pair with 4 juveniles in the extreme south-west on 19 July 1987 may possibly reflect local breeding, but the flock of about 100, including juveniles, in two tetrads in Linley Big Wood on 10 July 1990, a year with a major irruption, is unlikely to do so.

The earliest records are "a few pairs bred at Clive in 1913, 1915 and 1916, and in the Broseley/Wenlock area, possibly in 1928 and 1931", but those since then illustrate the Crossbill's irruptive nature. Influxes occurred in 1956, 1958 and 1962 and were followed by several suspected though unconfirmed breeding records in the following springs (*Handlist* 1964). Blank years recently between 1965 and 1970, and in 1974 and 1975, have been followed by invasions and records over a number of subsequent years.

Crossbill irruptions have increased in their scale since 1966 (*Pop. Trends* 1990), and this trend has been mirrored in Shropshire. The maturing plantations in upland regions may well allow the establishment of an annual breeding population, but it is unlikely that one existed during the Atlas period. It may never be possible to provide a population estimate for such an irruptive species.   *GT*

JS

BULLFINCH

Pyrrhula pyrrhula

Status: **Resident**

Tetrads with evidence of breeding

Confirmed	217 (25%)
Probable	359 (41%)
Possible	133 (15%)
Total	**709 (81%)**

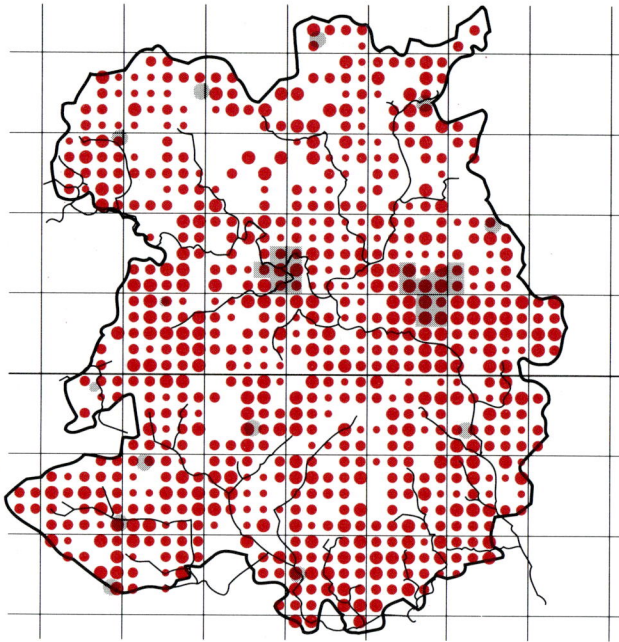

Bullfinch distribution broadly matches that of deciduous woodlands (Map 8), its preferred nesting habitat. Colonisation of secondary habitats such as copses and mature hedgerows has resulted in a fairly even distribution, with large gaps only on the open moors of The Long Mynd and the arable farmland of the northern plain.

Typical woodland seeds such as ash keys have been recorded as regular food items (*SBRs*) but the much publicised liking for buds of fruit flowers in early spring, though recorded, has never been a problem in a county with very few commercial orchards.

Pairs of Bullfinches stay in close contact when not actually incubating and are easy to locate by their calls. The resulting high incidence of probable breeding also reflects the difficulty of finding nests, as laying often starts relatively late in spring when vegetation is well grown, and of observing food for young in the beaks of seed-eating species. Proof is most likely to come from finding a family party with recently fledged young.

It was "considered to have increased in numbers in recent years" (*Handlist* 1964), and nationally CBC indices have shown that a high population in the 1960s and 1970s decreased in the early 1980s and stabilised at a lower level. This earlier growth followed by a recent downward trend has been attributed to the effect of predation by the declining, then growing, population of Sparrowhawks (*Pop. Trends* 1990). No research has been done on this topic in Shropshire.

The BTO *Atlas* (1976) reported average CBC densities in 1972 of 2.4 pairs per sq. km on farmland and 7.1 pairs per sq. km in the favoured deciduous woods. Since then CBC indices have demonstrated a 70% drop in population density on farmland and 40% in woodland. However, Shropshire has a high proportion of deciduous woodland compared to most English counties, and although the Bullfinch is never numerous, its widespread occurrence suggests a population of 1500–3000 pairs based on an assessment of around 2–4 pairs per tetrad.   *GT*

162

HAWFINCH
Coccothraustes coccothraustes

Status: **Resident**

Tetrads with evidence of breeding

Confirmed	5 (	1%)
Probable	6 (	1%)
Possible	3 (	0%)
Total	**14 (**	**2%)**

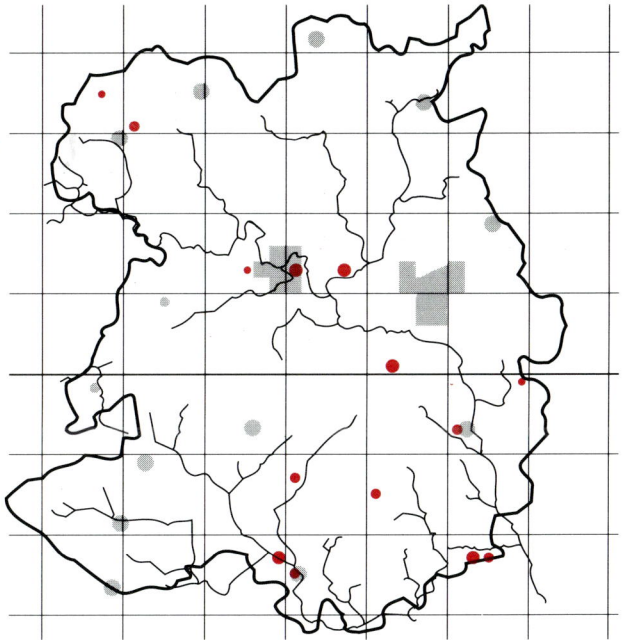

The Hawfinch is the largest resident finch, and extremely uncommon. Many records are of small winter flocks in favoured areas such as Whitcliffe, Ludlow. During the breeding season shy, elusive behaviour makes it much more difficult to detect, particularly when the tree canopy becomes fully developed.

Favoured habitats are deciduous woodland and mature parkland, especialy where hornbeam, wild cherry, beech or sycamore provide the large fruits and seeds on which it feeds.

The Hawfinch is most often located by its characteristic "tic" call, suggesting a rather loud Robin, heard during frenetic display flights. Proof of breeding is most likely to come from observing family parties not long after the young have left the nest.

The Hawfinch is sparsely distributed. In eight of the tetrads it was found in parkland or similar open areas with mature trees; whilst in four others it was found in mixed or deciduous woodland. Two records came from large overgrown gardens; one in Shrewsbury where fledged young were seen, the other in Bridgnorth where a pair were located.

"In 1899 H.E. Forrest considered that it had recently multiplied in numbers, having previously been reckoned as rare. It continued to be recorded as breeding until about twenty years ago in a number of widely scattered localities but had become "Resident, scarce. Infrequently recorded in recent years" with "Records suggesting breeding" in the 1950s from Leighton, Shirlett, Apley Castle and Bucknell (*Handlist* 1964). In the BTO *Atlas* (1976) almost all local records — 4 possibles, 1 probable and 3 confirmed breeding — came from 10-km squares in the south and south-east, but some of them are not repeated on this map. As the Hawfinch is sedentary and easily overlooked, with no obvious trend in population, these areas probably still hold breeding pairs, though numbers are thought to fluctuate and occupation of individual sites may be erratic (*Pop. Trends* 1990).

All records are likely to refer to nesting pairs, and it is certainly under-recorded. Assuming that Shropshire is typical of the country as a whole, a breeding population estimate can reasonably be based on an average of 10–20 pairs for each 10-km square where it was found (BTO *Atlas* 1976). The population may therefore be in the region of 110–220 pairs, though the shortage of records makes this even more of a "guesstimate" than usual and suggests that it may perhaps be too high.   *JM*

YELLOWHAMMER
Emberiza citrinella

Status: **Resident**

Tetrads with evidence of breeding

Confirmed	541 (62%)
Probable	269 (31%)
Possible	50 (6%)
Total	**860 (99%)**

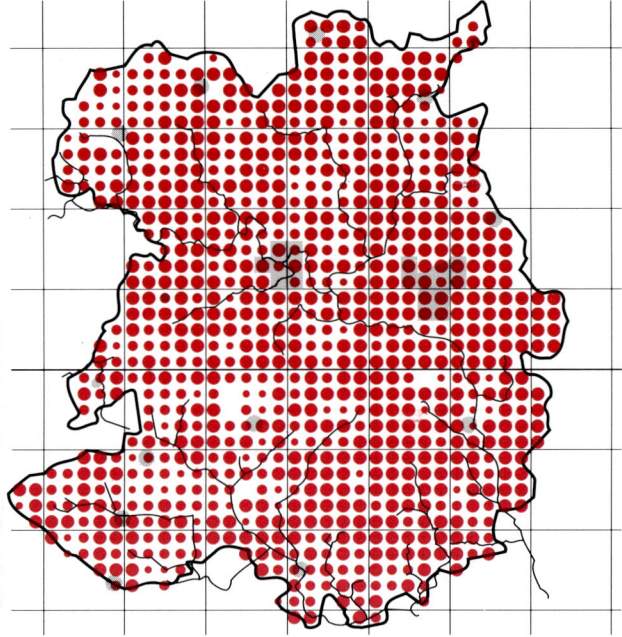

The Yellowhammer is very sedentary, ringed birds usually being recovered near the original ringing location.

Most open areas are inhabited provided there are bushes or shrubs to conceal the nest and song perches nearby. Farmland with mature hedgerows is the main habitat, but heaths, bracken-covered hillsides, commons and patches of scrub are also utilised.

The bright yellow plumage of the male is conspicuous as it sings "a little bit of bread and no cheese" from a prominent position, making territories easy to find. Pairs can be located early in the season, but females and young birds are much duller and are not so readily seen when the male is absent. Once nesting begins they become very secretive. Confirmation of breeding can be obtained when young are being fed, adults often collecting insects, grubs and caterpillars along the sides of country lanes.

Yellowhammers are distributed widely, being absent only from some urban areas and high open moorland in the southern hills.

The number of territories on a local CBC plot fluctuates slightly, but averages 10 pairs/sq. km (T.W. Edwards, pers. comm.), similar to the figure given in the BTO *Atlas* (1976), so the population is probably around 35,000 pairs.   *APD*

REED BUNTING
Emberiza schoeniclus

Status: **Resident**

Tetrads with evidence of breeding

Confirmed	91 (10%)
Probable	99 (11%)
Possible	90 (10%)
Total	**280 (32%)**

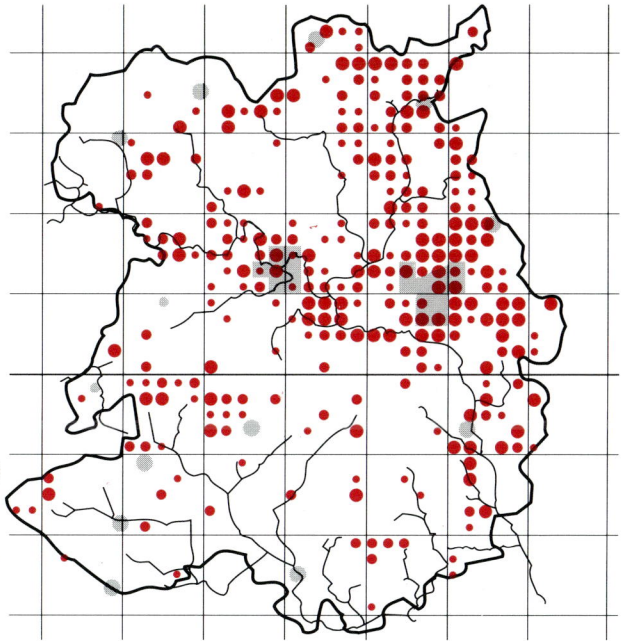

The Reed Bunting's preferred breeding habitat is marshy ground and wet areas with rank vegetation. The monotonous double whistle of the male's song enables territories to be located and confirmation of breeding can be obtained when young are being fed.

Many territories are located along the rivers Severn, Tern and Worfe, and their smaller tributaries, which drain the low-lying plains of the north-east. Reed Buntings are also found by pools and drainage ditches in the Telford area, mosses in the north and the Shropshire Union Canal, particularly the disused section between Welsh Frankton and Llanymynech.

West of the Severn valley the pattern changes considerably. Sluggish and still waters are less frequent, though breeding may occur where they do exist, such as at Shelve Pool at 300m and near The Bog at 365m. Fast-flowing rivers lack the marginal vegetation found by the slower waters and are unsuitable. Reed Buntings are found instead on high ground where rushy pastures and boggy areas are created by heavy rainfall. Titterstone Clee, Brown Clee and Clun Forest all have clusters of breeding sites, but the one near Pole Cottage close to the top of The Long Mynd at 470m is probably the highest.

During the Atlas period 25+ pairs were found on a WBS site on 4km of the River Severn between Atcham and Wroxeter, several broods were raised between Wroxeter and Buildwas, and 10 pairs were located at Chelmarsh (*SBRs*), but the density away from a few favoured sites is very low. This could account for the small proportion of tetrads with confirmed breeding records. Nationally the population during the survey period was at a low level (*Pop. Trends* 1990). Atlas fieldwork has found a wide variation in breeding densities, perhaps averaging 3–10 pairs per occupied tetrad, giving an estimated population of 570–1900 pairs. *APD*

CORN BUNTING
Miliaria calandra

Status: **Resident**

Tetrads with evidence of breeding

Confirmed	39 (4%)
Probable	91 (10%)
Possible	58 (7%)
Total	**188 (22%)**

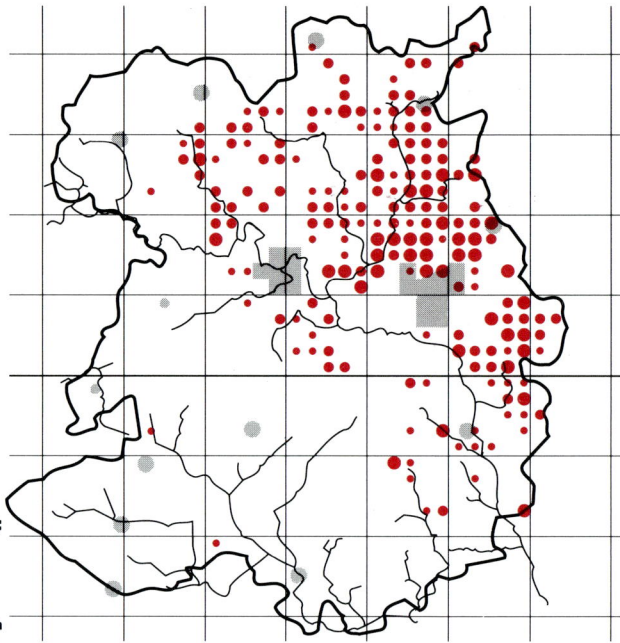

Male Corn Buntings establish breeding territories from April onwards when their short persistent jingling song is given from a wire, fence post, bush or other exposed position. Nests are usually in large arable fields so it is hard to see adults feeding young or to distinguish fledglings, making confirmation of breeding difficult.

Most Corn Buntings are found on the Shropshire plain between Telford and Market Drayton and the eastern sandstone plain between Bridgnorth and Shifnal. Further groups are scattered throughout the north extending as far as Rednal in the west. The Severn valley and a few smaller valleys in the south also produced records. The distribution of Corn Buntings is directly linked to the distribution of barley and these low-lying areas are intensively arable (see Map 7). Due to crop rotation breeding sites may change, and some tetrads near the edge of the range were not occupied every year, so the map probably overstates the position for any one year.

Although present over a wide area, they are not common. The highest numbers of singing males reported were from Lilleshall, Sutton Maddock and Baggy Moor, with up to 9, 12 and 5 respectively (*SBRs*). Elsewhere some tetrads appear to contain only one singing male. An average of 3–7 pairs per occupied tetrad gives an estimated population of between 400 and 900 pairs.

The success of the Corn Bunting has varied enormously as agriculture has changed. By 1940 it had disappeared from large areas of the country, then increased until the early 1970s when a decline began again, worsening in the 1980s (*Pop. Trends* 1990). Shropshire initially followed this pattern. "Though considered a widely but sparsely distributed breeding species at the turn of the century, it had become quite rare by 1940" and "in recent years . . . up to 10 pairs may be present" to the north-west of Wellington and in the area of The Weald

Moors, with odd birds elsewhere (*Handlist* 1964). Numbers gradually increased and spread (*SBRs*) and the BTO *Atlas* (1976) had records from 12 10-km squares. In contrast to the national trend this expansion has continued with Corn Buntings now being found in 10 additional 10-km squares, whereas the provisional distribution map for the New BTO *Atlas* (BTO *News* 173) shows an "enormous contraction in range . . . over the last twenty years".

Throughout the 1980s the number of singing males and the size of the winter flocks reported has been increasing (*SBR*s). Greater observer coverage may account for some additional records and wintering birds could be from elsewhere, but it appears that Corn Buntings are not declining and could still be increasing. However, their future is dependent on agricultural practice and in the present economic climate that is impossible to predict.   *APD*

MISCELLANEOUS SPECIES

A number of records were submitted on Atlas cards for species of which no map or account has been given. These species are all considered to be either winter visitors lingering into April or early May, passage migrants, summering non-breeders or escapes and feral birds; they are listed below. Unless specifically stated otherwise, all were recorded only as present or present in suitable breeding habitat, so no evidence of probable or confirmed breeding was obtained. As the non-breeding status of these species is well known, many observers will not have recorded them on Atlas cards even when they were seen. Any description of their status derived from Atlas records would therefore be misleading, and a more complete account of observations each year can be found in *SBRs*.

Miscellaneous Species recorded on Atlas Cards:

CORMORANT *Phalacrocorax carbo*
Winter visitor, small numbers in summer, mainly immatures.

SNOW GOOSE *Anas caerulescens*
Escapes or feral birds recorded several times each year. 5–8 seen frequently at Nib Heath from 1986 onwards were recorded as probably breeding, and suggest that a feral breeding population may become established in future.

EGYPTIAN GOOSE *Alopochen aegyptiacus*
Escape or feral, first listed in *SBR* in 1988.

WIGEON *Anas penelope*
Winter visitor, several large annual flocks, with small numbers sometimes remaining into April. Two breeding season records: a drake at Chelmarsh in May 1987 and a duck at Venus Pool through most of May and June 1990.

PINTAIL *Anas acuta*
Winter visitor but a pair remained at Polemere into early June 1987.

RED KITE *Milvus milvus*
Recorded in five of the Atlas years, with a total of 11 breeding season records, mainly in early April or July.

GOLDEN PHEASANT *Chrysolophus pictus*
Rare feral or escape.

GOLDEN PLOVER *Pluvialis apricaria*
Winter visitor, often in large flocks.

GREEN SANDPIPER *Tringa ochropus*
Mainly a passage migrant, also a wintering visitor, but May/June records received 1986–90, usually singles.

LESSER BLACK-BACKED GULL *Larus fuscus*
Common winter visitor, some summering non-breeders, mainly immatures.

HERRING GULL *Larus argentatus*
As Lesser Black-backed Gull.

COMMON TERN *Sterna hirundo*
Annual passage migrant, usually small numbers.

LITTLE TERN *Sterna albifrons*
Only five records this century before 1987, then a few passage records of one or two each year to 1989.

SHORT-EARED OWL *Asio flammeus*
Annual winter visitor in small numbers but a June record in both 1987 and 1989.

WRYNECK *Jynx torquilla*
Now believed to be only a passage migrant (see next section on species no longer breeding). Recorded annually since 1984, mainly on return passage, with spring records of singles in April 1985, May 1987, April 1988 and May 1990.

FIELDFARE *Turdus pilaris*
Common winter visitor, often in large flocks until mid-April, generally departing late April, with individuals occasionally lingering into May.

BRAMBLING *Fringilla montifringilla*
Annual passage migrant and winter visitor, occasionally in large flocks. Small numbers may linger into April or early May.

Of these species, only Golden Plover and Wryneck are definitely known to have bred in Shropshire this century. Many others, known to be regular passage migrants, were also recorded in April and May, as listed in *SBR*s, but as these records were not submitted on Atlas cards, and there is no evidence of breeding beyond presence, they are not included in the above list.   LS

SPECIES NO LONGER BREEDING

Several species which have bred in the last 40 years — Black Grouse, Corncrake, Nightjar, Wryneck, Woodlark, Black Redstart and Red-backed Shrike — were not proved breeding during the Atlas period. All except Black Redstart, and perhaps Black Grouse, previously bred annually.

Atlas fieldwork did locate Corncrake, Nightjar and Black Redstart, and they are included in the species accounts. Of the remainder only Wryneck produced any Atlas records. The recent history of these species is summarised below.

BLACK GROUSE *Tetrao tetrix*
Sedentary, they require a habitat mosaic, usually of heather moorland, woodland and bog, to meet the nesting and feeding needs of adults and young throughout the year. Loss of this habitat has caused a serious decline over the last 100 years and they are now absent from much of England, though they have recovered somewhat in Wales by taking advantage of young conifer plantations (*Red Data* 1990).

The *Handlist* (1964) described Black Grouse as "resident, very scarce" with unconfirmed reports of recent sightings from The Long Mynd and Stiperstones, and plantations at Cefn Coch near the Welsh border in the north-west. "It was last reported to have bred on Titterhill, Bucknell, in 1954, but there are no recent records from here or Clun Forest, where it occurred near the Radnor border".

The BTO *Atlas* (1976) recorded it present in SO39, though not all this 10-km square is in Shropshire, and there are *SBR* records of a male in the north-west in 1970, a bird on The Long Mynd "possibly introduced by keepers" in 1975 and a female there in 1976, and the last record, in 1981, of a "Grey-hen reported near Earl's Hill N.R. Believed to be of introduced origin".

WRYNECK *Jynx torquilla*

A summer visitor, formerly widespread over England and Wales, it has suffered a massive decline over the last 150 years, although since 1969 a few pairs have been present in the Scottish Highlands. Cooler, wetter summers, which reduce the availability of ants, the staple diet, are thought to be responsible (*Red Data* 1990). The main habitat requirements are open areas with trees that provide holes for nesting and short grass for feeding.

The *Handlist* (1964) described the Wryneck as extremely rare, though "in the early years of this century breeding occurred regularly in the southern part of the county". At known regular sites nesting was last recorded at Broseley (1920), Dowles (1924) and Bucknell (1952 and 1953), the latter being the most recent confirmed breeding, though it was seen at all these sites for one or more years subsequently. The only other recent record is of a nest at Edgerley near Nesscliffe in 1941.

The BTO *Atlas* (1976) was blank, but 1973 produced two different records, each of a Wryneck being seen at the same place twice more than a week apart in May and June, perhaps indicating possible breeding, and one was seen in one of these two areas in June the following year. In the 10-year period 1975–84 there were nine records, all except two in August or September. One of these from Wenlock Edge in mid-July 1980 was the last record outside the passage period. During the Atlas years it has been recorded every year, but only four times in the breeding season, all of individuals during the spring passage period.

WOODLARK *Lullula arborea*

Formerly widespread and numerous, it is now rare with less than 230 pairs breeding in Britain. It inhabits lowland heaths and other open areas with scattered trees or shrubs for song posts, long grasses for nesting and short vegetation or bare ground for feeding. Habitat loss coupled with hard winters is believed to be the cause of the reduction in population and range (*Red Data* 1990).

The *Handlist* (1964) described it as "resident, local in small numbers in the south and west of the county . . . Numbers appear to fluctuate from year to year but it is probably overlooked". However, no *SBR* records were received in 1964, and in 1965 a few Woodlarks were reported from the Bucknell area "otherwise several areas where previously found, searched but none located". The only subsequent records came from Black Rhadley Hill in May 1976, and a singing male at Red Wood near Clun in June 1972, apart from a road victim near Shifnal in February 1985 and an individual found exhausted near Wellington at the end of July 1987 which was released the following day. The BTO *Atlas* (1976) showed it present in SJ22, but not all this 10-km square is in Shropshire. The last record of confirmed breeding was "at the Ercall near Wellington, July 29, a pair with young" in 1957 (*SBRs*).

RED-BACKED SHRIKE *Lanius collurio*

Formerly widespread in England and Wales, it suffered a marked decline from the mid 19th century and by the 1980s was restricted to East Anglia. Only one or two pairs nested there in 1987 and 1988, and none at all in 1989, so it is now almost certainly lost as a breeding species in Britain. The reasons for the decline are not understood, but may be due to climatic change affecting the food supply, mainly grasshoppers, butterflies, moths, bees and beetles (*Red Data* 1990).

It used to inhabit commons, waste ground, scrub and overgrown hedgerows, young plantations and especially lowland heath. The *Handlist* (1964) described it as a "breeding visitor, now very rare. Formerly not uncommon in suitable habitats throughout the county. The last definite records of breeding were of a pair at Bucknell in 1954 and in 1946 when the late J.H. Owen found three pairs nesting on Llynclys Hill". His studies there since 1881 had found as many as 15 pairs, with 5 in 1940 rising to 9 in 1944. Several other main breeding areas were also noted.

The only records since 1954 are of single males near Whitchurch in June 1981 and near Leebotwood in July/August 1981, a juvenile in November 1987 at Lower Wood, and another male in Cardingmill Valley on The Long Mynd in June 1988. These last three records came from within four miles of each other.

Several other species are known to have bred in Shropshire previously, though none of these has been confirmed since 1914, and they are included in Table 1. Of those recorded breeding since Forrest's day at the turn of the century, Golden Plover has bred only three times, twice in the early years of the century and then a nest and eggs were found on The Long Mynd in 1976 (*SBR*), and Cirl Bunting "was never common, but more frequently recorded prior to 1920" with the last breeding season record in 1956 (*Handlist* 1964).

Table 1. Species no longer breeding in Shropshire

Lost since 1950			Last bred before 1915	
Species	Previous status	Last confirmed breeding date	Species	Last confirmed breeding date
Black Grouse	A	1954	Bittern	1836
Corncrake	A	1975	Garganey	1888
Golden Plover	3	1976	Pintail	1881
Nightjar	A	1983	Red Kite	19th C.
Wryneck	A	1953	Honey Buzzard	19th C.
Woodlark	A	1957	Green Sandpiper	1888
Black Redstart	3	1978	Cirl Bunting	1904
Red-backed Shrike	A	1954		

Status: A = Annual.
Number: The total number of known confirmed breeding records.

LS

SUMMARY AND CONCLUSIONS

The Atlas represents the greatest contribution to our knowledge of the county's breeding birds since the formation of the Shropshire Ornithological Society in 1954. It has clearly demonstrated the dependence of most species on specific habitats, particularly by showing just how few species are so adaptable that they can breed in every tetrad. Even some of the most widespread and numerous species are missing from a few tetrads, or they are scarce enough to be overlooked as their habitat is in such short supply. Perhaps the most surprising of all the results of the Atlas is that these really common species do not breed throughout the county.

Table 2: Status of widespread and numerous species

Species believed to breed in every tetrad	Species which, perhaps surprisingly, do not appear to breed in every tetrad	
	Species	**Absent from tetrads consisting primarily of:**
Woodpigeon	Pheasant	Urban areas, high ground without copses.
Wren	Swallow	Moorland with no buildings.
Dunnock	House Martin	Sparsely populated upland, intensive arable and woodland.
Robin		
Blackbird	Song Thrush	Upland, intensive arable.
Blue Tit	Willow Warbler	Intensive arable.
Magpie	Great Tit	Intensive arable.
Carrion Crow	Jackdaw	Urban areas, moorland, intensive arable.
Chaffinch	Starling	High open moorland.
	House Sparrow	Moorland with no buildings.
	Yellowhammer	Urban areas, high open moorland.

Note: Species included in this table were recorded in over 90% of tetrads with probable or confirmed breeding established in over 80% of tetrads.

Most species are so dependent on specific habitats that changing land-use may have a major effect on their population level, range and distribution. Climate is another factor, including that on the migration routes and in the wintering areas of our summer visitors. In a few cases conservation measures have allowed rare species to expand from protected refuges. These factors often give rise to an underlying trend, sometimes strong, of growth or decline, even if the breeding success and winter survival rate of many birds fluctuate from year to year. A common strand through many of the accounts of resident species typical of farmland is a marked decline in numbers due to greater use of herbicides and pesticides, and autumn sowing of cereal crops reducing food supplies. The population of many other species has grown due to a reduction in the number of gamekeepers, increased legislative protection, or creation of new man-made habitats such as gravel pits. The increase in woodpeckers and Nuthatch, attributed to Dutch elm disease, may be temporary as the resulting dead wood will eventually disappear.

For some species the population trend in Shropshire can be determined from local evidence, by comparing the current map with that in the BTO *Atlas* (1976) or with information in the *Handlist* (1964) or *SBR*s. In other cases the national changes outlined in *Pop. Trends* (1990) are likely to apply locally, but quantitative evidence to be absolutely certain is currently lacking.

The changing status of species now breeding in the county is summarised in Tables 3 (overleaf) and 4, and those no longer breeding are summarised in Table 1 in the previous section. A full explanation of any changes is given in the individual species accounts.

Table 4. Species thought to have changed status over the last 30 years as national trends are believed to apply in Shropshire

Increased	Recovering from recent decline	Decreased	
Coot	Stock Dove*	Kestrel	Spotted Flycatcher
Blackcap	Meadow Pipit	Lapwing	Marsh Tit
Coal Tit	Garden Warbler*	Barn Owl	Starling
Jay	Chiffchaff	Skylark	House Sparrow
Jackdaw	Greenfinch	Swallow	Tree Sparrow
Carrion Crow	Goldfinch*	Dunnock	Linnet
Chaffinch		Blackbird	Bullfinch
		Mistle Thrush	Reed Bunting

* Population still well below earlier level.

For many species not listed in the tables there is no evidence of change, though of course it may be happening undetected. Several resident species, including Kingfisher, Wren, Goldcrest and Long-tailed Tit, are omitted as any underlying trend is obscured by the considerable fluctuations in population due to winter weather conditions. Dipper is also omitted as the substantial decrease at the end of the Atlas period is considered to be due to abnormally dry summers. Some species appear to be decreasing as a result of both cold winters and changing farm practice, for example Song Thrush and Reed Bunting, and these are included in the table to highlight the effect of habitat changes on population levels.

The Atlas has also contributed to the knowledge of individual species. Though some of those breeding for the first time would have been discovered anyway, as they are conspicuous, inhabit well-watched nature reserves or were first reported by people not engaged on fieldwork, Mandarin and Goosander were found by Atlas workers. New Goshawk and Hobby nests were located, and a previously unsuspected Merlin site was discovered. Many areas in the south and west are remote and rarely visited by birdwatchers, so the systematic coverage provided by the Atlas is the first time an overview has been obtained for many species, especially those of the uplands. Tufted Duck, Red-legged Partridge and Reed Bunting have been found in upland habitats where they would not have been expected from the commentary in the BTO *Atlas* (1976). Farmland is also rarely

Table 3: Species with changed status in Shropshire over the last 30–40 years, clearly demonstrated by local evidence (see Table 1 for those no longer breeding)

Bred for the first time (year)	Increased contrary to national trend	Increased		Recovery from recent decline	Decreased
Black-necked Grebe 1985	Grey Wagtail	Great Crested Grebe	Great Spotted Woodpecker	Grey Heron	Red Grouse
Greylag Goose 1969	Stonechat	Canada Goose	Lesser Spotted Woodpecker	Mute Swan*	Grey Partridge
Barnacle Goose 1984	Corn Bunting	Tufted Duck	Lesser Whitethroat	Sparrowhawk	Snipe
Shelduck 1963		Goshawk	Pied Flycatcher	Sand Martin**	Redshank
Mandarin 1988		Buzzard	Nuthatch	Redstart	Turtle Dove
Gadwall 1980		Merlin (?)	Magpie	Sedge Warbler*	Nightingale
Goosander 1987		Hobby	Raven	Whitethroat*	Whinchat
Ruddy Duck 1965		Red-legged Partridge	Siskin	Rook*	Wheatear
Marsh Harrier 1988		Feral Pigeon	Common Crossbill		Song Thrush
Peregrine 1970		Little Owl			Grasshopper Warbler
Oystercatcher 1981		Green Woodpecker			Redpoll
Little Ringed Plover 1976					
Collared Dove 1963					

* Population still below earlier level. ** Population still far below earlier level.

studied, and the distribution of Yellow Wagtail and Corn Bunting might not have been predicted before the Atlas started. As an incentive to further research, there must be some more Hobby nests somewhere!

Finally, the Atlas has identified a number of other topics that require further study. One of the most intriguing is the marked easterly bias in the Kestrel's distribution. Another is the apparent preponderance of Garden Warbler over Blackcap, and Willow Tit over Marsh Tit, in the south-west, whereas the converse is true in the county as a whole, as summarised in Table 5.

Table 5. Distribution of Garden Warbler and Blackcap, and Marsh and Willow Tit, comparing the south-west with the county as a whole.

Occupied by	Whole county 870 tetrads		South-west 110 tetrads	
	No.	%	No.	%
Garden Warbler — total	461	53.0	77	70.0
Blackcap — total	646	74.3	68	61.8
Garden Warbler (not Blackcap)	67	7.7	24	21.8
Blackcap (not Garden Warbler)	252	30.0	15	13.6
Willow Tit — total	223	25.6	40	36.4
Marsh Tit — total	290	33.3	34	30.9
Willow Tit (not Marsh Tit)	112	12.9	24	21.9
Marsh Tit (not Willow Tit)	179	20.6	18	16.4

Similar comparisons might apply to other closely related species but the Atlas fieldwork method is not sufficiently detailed to demonstrate it.

All species have been under-recorded to some extent, owing to the fieldwork method and the large area to be covered, but some species were not found at places where they might have been expected, for example, Little Grebe, Sparrowhawk and Curlew. Other species at the edge of their range or otherwise vulnerable, for example some of the ducks and raptors, Red Grouse, the partridges, Oystercatcher, Redshank, Snipe, Little Ringed Plover, Turtle Dove, Nightjar, Nightingale, Whinchat, Stonechat, Ring Ouzel, Reed Warbler, Wood Warbler, Firecrest, Raven, Tree Sparrow, Redpoll, Hawfinch and Corn Bunting, merit further research.

The same applies to the numerous and widespread species not found in every tetrad, listed in Table 2. Were they overlooked, or are they genuinely absent? Of the scarcer but expanding species, have Barnacle Goose, Shelduck, Mandarin, Gadwall, Shoveler, Goosander, Siskin and Common Crossbill established annual breeding populations?

The population estimates are often the result of informed guesswork, and they should be revised as more thorough assessments are made. One of the major lessons of the Atlas is the need for more detailed records to be submitted by observers for the *Shropshire Bird Report* (*SBR*). In trying to assess the current population and establish whether there has been any change in numbers or distribution, the most useful *SBR* records were those that gave counts, especially of singing males or pairs; a precise location, so the record could be related to one or more specific tetrads; and an indication of any evidence of breeding, particularly whether two birds were a pair, or parties included recently fledged young or juveniles. Hopefully the experience gained from fieldwork will encourage more useful records in future. Records for species in tetrads where they are missing in the Atlas or that improve the level of evidence of breeding, will be especially welcome.

Every effort will be made to update the information in the Atlas each year in the *SBR*, though this may only be possible for the more scarce species, except when specific surveys are organised. However, the Atlas Project has clearly achieved its original aim of providing a benchmark against which future trends can be assessed.

Atlas fieldwork has also identified important sites, and these are now being surveyed so they can be incorporated into SWT's Register of Prime Sites for Nature Conservation. To co-ordinate this work and build upon the results of the Atlas the SOS has established a Conservation Committee to organise future fieldwork activities, involve the members and seek to answer some of the many questions highlighted by the species accounts and interpretation of the maps. The Atlas Sub-Committee will be making recommendations, but any proposals for future fieldwork, and offers of help, will also be gratefully received.

The Atlas has taught us all how little we know about Shropshire's breeding birds, and the Society intends to build upon the results to continually increase and update our knowledge and understanding of this important topic. *LS*

APPENDIX 1. Fieldwork and Results

SURVEY AREA AND MAP REFERENCES

To create manageable survey areas, and in common with similar atlases elsewhere, Shropshire was divided into 2-km squares, known as "tetrads", using the National Grid printed on all Ordnance Survey (OS) maps. On the 1:50,000 series OS maps the grid lines are shown as pale blue and define 10-km and 1-km squares. Each 10-km square has a unique number, and those in Shropshire are shown on Map 18.

Every 10-km square is then further divided into 25 tetrads, each with a separate letter taken from the tetrad letter key shown below.

Each tetrad therefore has a unique number, consisting of the 10-km square number, followed by the tetrad letter. For example, Shrewsbury Castle is in tetrad SJ41W.

At the edges of the county, tetrads were included in the Atlas if more than half of their area is in Shropshire. In such cases the whole tetrad was surveyed, not just the part

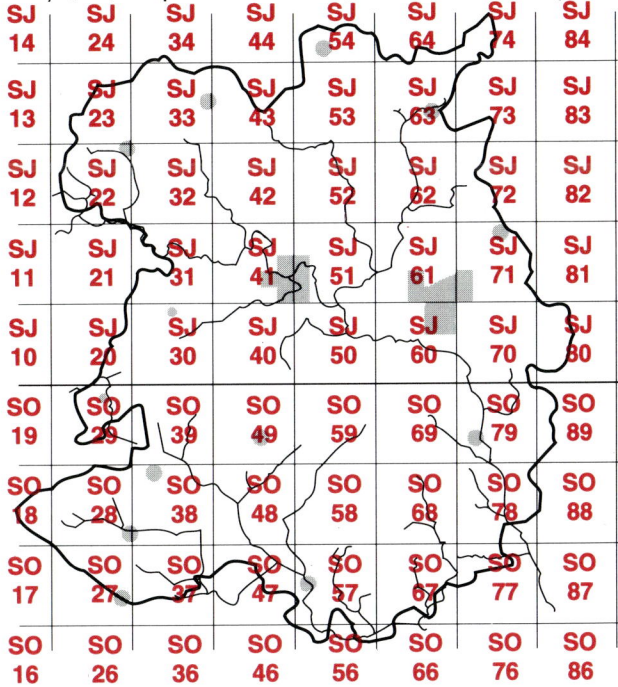

SJ 14	SJ 24	SJ 34	SJ 44	SJ 54	SJ 64	SJ 74	SJ 84
SJ 13	SJ 23	SJ 33	SJ 43	SJ 53	SJ 63	SJ 73	SJ 83
SJ 12	SJ 22	SJ 32	SJ 42	SJ 52	SJ 62	SJ 72	SJ 82
SJ 11	SJ 21	SJ 31	SJ 41	SJ 51	SJ 61	SJ 71	SJ 81
SJ 10	SJ 20	SJ 30	SJ 40	SJ 50	SJ 60	SJ 70	SJ 80
SO 19	SO 29	SO 39	SO 49	SO 59	SO 69	SO 79	SO 89
SO 18	SO 28	SO 38	SO 48	SO 58	SO 68	SO 78	SO 88
SO 17	SO 27	SO 37	SO 47	SO 57	SO 67	SO 77	SO 87
SO 16	SO 26	SO 36	SO 46	SO 56	SO 66	SO 76	SO 86

Map 18. Shropshire's 10-km squares.

inside the county boundary. If more than half the tetrad is in the neighbouring county, it was excluded. Thus a few records will have come from just outside Shropshire, and small fragments at the edge of the county have not been covered.

Shropshire is England's largest landlocked county, and the survey covered 870 tetrads (3480 sq. km). A separate survey was carried out for each tetrad, and a dot on a distribution map means that species was recorded in the appropriate habitat somewhere within the 2-km square at that position on the map.

TETRAD LETTER KEY

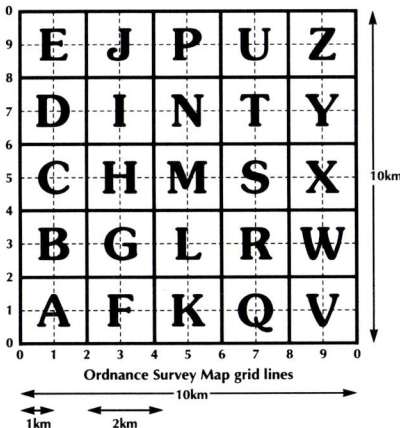

E	J	P	U	Z
D	I	N	T	Y
C	H	M	S	X
B	G	L	R	W
A	F	K	Q	V

Ordnance Survey Map grid lines

10km

1km   2km

The species accounts frequently refer to places where particular birds were found. Appendix 5 lists every place mentioned in the accounts, with a four-figure grid reference which identifies the 1-km square on the map where that place can be located. The first two numbers in the grid reference are those printed at the top and bottom of the map to define lines that run north to south, known as "eastings"; and the second two numbers are those printed on each side of the map to define lines that run east to west, known as "northings". Where the two lines defined in the grid reference intersect is the bottom left-hand corner of the appropriate 1-km square. As the 1-km grid lines divided each 10-km square into 100 1-km squares, the number of the 10-km square is the first and third number in the four-figure grid reference of the 1-km square. For example, Shrewsbury Castle is immediately to the east of "easting" 49, and to the north of "northing" 12, so its four-figure grid reference is SJ4912, in 10-km square SJ41.

RECORDING THE EVIDENCE OF BREEDING
In 1979 the European Ornithological Atlas Committee established a standard of 17 categories of breeding evidence grouped into four main grades or classes (present, possibly breeding, probably breeding and proven or confirmed breeding). The BTO has adopted a series of letter codes which correspond with these 17 categories as shown in Table 6, and they were used for this Atlas.

FIELDWORK
The county was divided into ten areas, each with an Area Co-ordinator to organise the fieldwork. Detailed "Instructions for Fieldworkers" were produced, which included guidance on defining the tetrad boundaries, so records could be mapped accurately. Printed record cards listing all the species likely to be found in the county, with three columns for the categories of breeding evidence, were also supplied to all fieldworkers.

Volunteers, mainly members of the SOS or SWT, were initially asked to survey one or more of their local tetrads. They were advised to identify all the different habitats using an OS map (preferably 1:25,000 scale) and visit them all during the breeding season, at least twice and several times if possible. Visits were also requested at dusk and night to find the crepuscular and nocturnal species.

The Instructions also encouraged fieldworkers covering an unfamiliar tetrad to seek out knowledgeable local people and accept records from them if they were certain the species and location had been correctly identified.

The main breeding season was defined as 1 April to 31 July, though earlier visits were requested for Crossbill, owls and Mistle Thrush, and later ones for pigeons, doves and Corn Bunting. Where appropriate, early records were also accepted for other resident early breeders such as Raven, and late records were accepted for species that raise second and third broods, and might still have young in the nest as late as August or September.

Proof of breeding was only required once for each species in each tetrad at any time over the whole Atlas period. No attempt was made either to establish population levels, or check presence every year. Results were submitted at the end of each season on the printed record card, and the Area Co-ordinator incorporated them onto a master card for each tetrad.

Table 6. Categories of Breeding Evidence

Class	Code 1	Code 2	Code 3	Category of Evidence
1. Present	✓			Species observed in breeding season.
2. Possibly Breeding	H			Species observed in breeding season in possible nesting HABITAT.
	S			SINGING male(s) present or breeding calls heard in breeding season.
3. Probably Breeding		P		PAIR observed in suitable nesting habitat in breeding season.
		T		Permanent TERRITORY presumed through registration of territorial behaviour (song etc.) on at least two different days, a week apart, at the same place.
		D		DISPLAY and courtship.
		N		Visiting probable NEST SITE.
		A		AGITATED behaviour or anxiety calls from adults.
		I		Brood patch on adult examined in hand indicating probable INCUBATING.
		B		BUILDING nest or excavating nest hole.
4. Confirmed Breeding			DD	DISTRACTION DISPLAY or injury feigning.
			UN	USED NEST or egg shells found (occupied or laid within the survey period).
			FL	Recently FLEDGED young (nidicolous species) or downy young (nidifugous species).
			ON	Adults entering or leaving nest site in circumstances indicating OCCUPIED NEST (including high nests or nest holes, the contents of which cannot be seen) or adult sitting on a nest.
			FY	Adults carrying FOOD for YOUNG, or faecal sacs.
			NE	NEST containing EGGS.
			NY	NEST with YOUNG seen or heard.

THE COMPUTER PROGRAM TETRAD™

Creating and maintaining distribution maps manually for over 120 species from 870 different master cards would have been extremely time-consuming and prone to error. A computer program was therefore written specially for the project. It is now marketed under the name TETRAD™ and details are given in the acknowledgements in Appendix 7.

For each tetrad, the observations recorded on every card were checked at the end of each season, and the highest level of breeding evidence obtained for each species was fed into the computer. At the end of the project the computer printout for each area was reconciled with the master cards held by the Area Co-ordinator, providing a valuable check against transcription errors from the record cards to either the master cards or the computer file.

As well as producing the results in map form, TETRAD™ will also produce them in list form, or produce maps or lists of tetrads where species are absent. It will also produce maps or lists combining up to any nine species, to compare distribution, and species lists for tetrads. Several of the maps in the section on "Shropshire and its Bird Habitats" were produced in this way.

PROGRESS

Area Co-ordinators attempted to persuade fieldworkers to cover different tetrads in the second and subsequent years, to ensure total coverage. New fieldworkers were also recruited, and training meetings were held each year. While some workers covered several new tetrads each year over the whole period, others were reluctant to do more than one or two home squares.

After the second year it was clear that the original plan for a three-year project was unrealistic, and a five-year target was set. After three years there were still 113 tetrads with no records at all. Most of these were in the west and south, reflecting the low human population levels in those areas, and consequent absence of fieldworkers.

At the end of the fourth year the coverage was analysed in detail. While only 20 tetrads still had no records, a further 159 were poorly covered, with less than 15 species recorded as probable or confirmed breeding, and less than 30 in total. The Committee set targets at this stage: "It should be possible to find at least 50 species in most tetrads, and establish probable or confirmed breeding for at least 30 species." In addition an acceptable level was set (40 species, with half or more as probable or confirmed breeding) and also a minimum level (30 species, with half or more as probable or confirmed breeding). The Atlas period was also extended for a further, sixth, season, to incorporate the additional records anticipated from fieldwork for the New BTO *National Atlas*, then planned to end in 1990.

At the start of the sixth and last year, 1990, there were 43 tetrads below the minimum level, and a further 71 below the acceptable level. Most of those below the minimum level were in the extreme west, and fieldworkers travelled from east and south-east Shropshire to cover many of them.

FINAL COVERAGE

At the end of the Atlas fieldwork only 3 tetrads were below the minimum target set by the Committee (SO49H, SJ72F and SJ73F), but all of these had more than 10 species recorded as probable or confirmed breeding, and more than 20 in total. A further 12 tetrads were below the acceptable level, with 133 below the target level.

The final levels of coverage achieved are shown in Maps 19 and 20 (opposite). Map 19 shows coverage in terms of the Committee's targets, and Map 20 shows the number of species for which breeding was confirmed.

BIAS IN THE RESULTS

Several factors influence the results obtained through this type of survey, involving many people over several years.

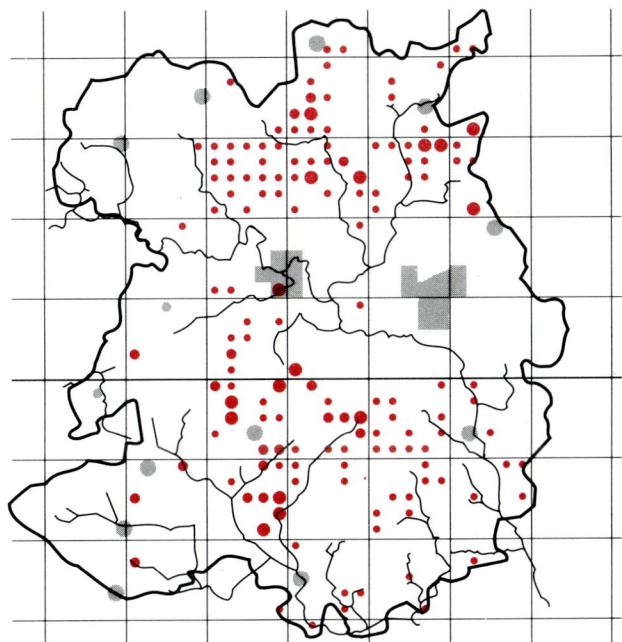

Map 19. Fieldwork coverage below the Committee's targets.

● Below the acceptable level (less than 40 species and/or less than 20 probable or confirmed breeding). Of these SO49H, SJ72F & SJ73F were below the minimum level.

● Below the target level (less than 50 species and/or less than 30 probable or confirmed breeding).

• Below the target level of 50 species but more than 30 probable or confirmed breeding.

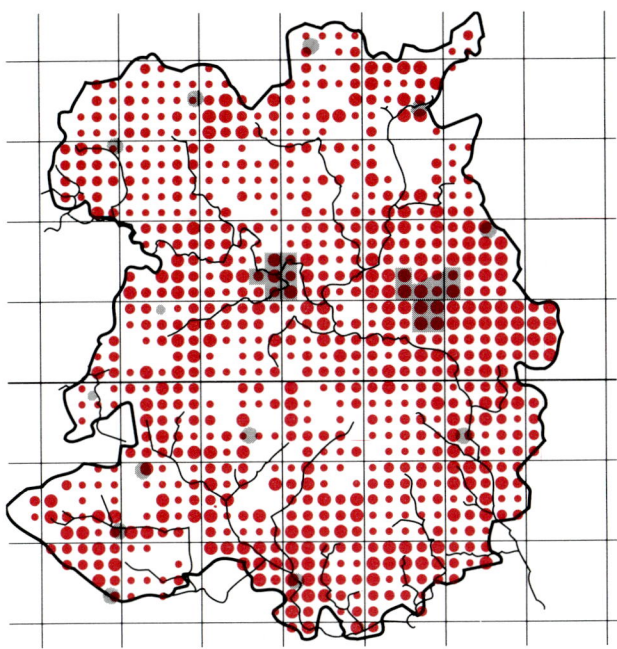

Map 20. Confirmed Breeding — total number of species.

● 40 or more.
● 30 or more.
• 20 or more.

Uneven fieldwork coverage

There is a tendency for the maps to reflect the distribution of observers and timing of their fieldwork visits, rather than the distribution of the birds. Almost half the population lives in Telford or Shrewsbury. Of the other towns only Bridgnorth and Oswestry have more than 10,000 people, and only 21 other towns have over 2000. The remaining population is scattered in small villages and isolated farms. There are many more people in east Shropshire than south, west or north. Not surprisingly, the distribution of fieldworkers largely mirrors the pattern of human settlement. Though 382 people contributed record cards to the Atlas, most helpers covered one or two local squares for part of the survey period, often in great depth. Though the Atlas could not have been completed without this help, the bulk of the fieldwork was done by around 50 experienced active birdwatchers, many of whom travelled widely. With 870 tetrads, many have not received thorough coverage.

Tetrads in the more remote parts of the county, especially in the hills to the west and south, and the heavily agricultural areas in the north, may have been visited by only one observer in only one season. Even the most skilled and dedicated observers are not going to find every species in just a few visits. Such visits are unlikely to be made during the evening or night, particularly if a long journey is necessary, and if they happen to be made in a bad season, either in general or for specific species, then the record will be even less complete. The more frequently a tetrad is visited, the more comprehensive the final record will be. It is interesting that very few record cards were received, even for well-covered squares, that did not include at least some improvement to previous records.

Visits from different people also improve results. Fieldworkers survey tetrads in different ways and at different times. Holidays at the end of the season may prevent a visit at the best time to confirm breeding of many species, but a late visitor will hear little song and might find access to the only pool prevented by overgrown nettles and brambles. People also have preferences for particular habitats, and different recognition skills, especially for the secretive small active species best identified by song. Some observers may even overlook the extremely common species, which they consider uninteresting.

Though most tetrads have been covered reasonably well, comparison of the final coverage maps (Maps 19 and 20) with those of each individual species shows that some of the gaps are due to poor coverage. This is particularly true for the very widespread species that are believed to breed in every tetrad.

Taking the coverage maps and fieldwork method together, it is clear that most species are under-recorded to some extent, but this is less likely to occur in the north-eastern quarter of the county.

Mapping error

Though fieldworkers were given clear guidance in identifying tetrad boundaries, a few records will inevitably be wrongly mapped either because the tetrad was named incorrectly on the record card or the fieldworker unknowingly crossed the boundary. Though the Area Co-ordinators trapped most of these errors, some will have escaped.

The birds too are mobile, and do not remain in their "home" tetrad. Raptors carry prey great distances to their nestlings, water birds may nest some distance from the pools where they eventually raise their young, and fledglings too may cross tetrad boundaries once family parties leave the nest site. Even pairs with small territories may occupy part of two tetrads, and breed in both in different years during the Atlas period. Fieldworkers were advised to record such observations in the tetrads in which they were seen, rather than guess where the nest may have been.

A few errors will arise from incorrect identification in the field, particularly of Marsh and Willow Tits and, to a lesser extent, Willow Warbler and Chiffchaff.

In spite of thorough checks, it is unlikely that transcription errors from the fieldworkers' record cards to the computer file have been totally eliminated.

Many records were submitted to the County Bird Recorder by observers not directly engaged in Atlas fieldwork. For the less common species, a check of the County Bird Report was carried out to ensure that there were no SBR records missing from the appropriate Atlas map. Many such records did not include an accurate map reference, so they have been included in the Atlas map at the most likely location, but again a few of them may not be in the correct tetrad.

All original fieldwork record cards, each attributed to its observer, are stored at the offices of the SWT.

Though the above factors may mean that a few dots should have been mapped in nearby tetrads, and the occasional one perhaps omitted, they are more than offset by the under-recording inherent in the fieldwork. They should not affect the broad interpretation of the maps.

The effect of weather

The weather influences results in two ways. First, more fieldworkers will be involved, and each will make more field trips, in good summers. The first four summers were relatively cold and wet, and only the last two were hot and dry.

Second, the ease of obtaining records is affected. The vast majority of confirmed breeding records arise from seeing either adults carrying food to their nest, or fledglings with their parents. Weather that reduces the food supply or the number of young raised for any particular species will reduce the chance of observers proving breeding that year. The number of breeding pairs of some resident species is reduced, perhaps considerably, if the preceding winter was hard, especially if there are several bad winters in succession. Also, the number of young birds raised depends on the abundance of the right food supply. Insect numbers will be reduced by cold springs or summers, their breeding cycle may be delayed, and they may spend less time on the wing, making it harder for insect-eating birds to gather enough of them.

Hot summers may dry out the ponds or mud in which various insects breed, reduce the water level of streams, ponds and marshes, and make the ground much harder, affecting the food supply or breeding habitat of other birds. This effect is exaggerated if the preceding winter and spring have been dry.

Heavy rain may damage some nests, the consequent floods may wash away nests at the water's edge, and if adults become wet while collecting food for nestlings they will carry moisture into the nest and the bedraggled young may die of cold.

While annual fluctuations due to the weather should not have affected results in tetrads which were visited several times in each year of the survey, they may have a significant effect on finding particular species in tetrads only visited a few times in one year. For example, an experienced fieldworker spending three full days in a tetrad will certainly achieve excellent results by the yardstick of Map 19. However, suitable habitats for Kingfisher or Dipper may only be visited for say an hour on each of the three days. As indicated by the species accounts, such restricted visiting time would be much more likely to find Kingfisher in the last two years of the survey than the first two. The hard winters of 1984–85 and 1985–86 reduced the population, but it recovered following three very mild winters. Hot, dry summers in 1989 and 1990 reduced water levels and turbulence, so catching fish was relatively easy, and two or three large broods could be raised. The opposite is true for the Dipper: reduced water levels and turbulence reduced the abundance of food in the water, and made nests more vulnerable to predation, so the population was at a very low ebb in 1990. Both species could breed successfully on the same stretch of river in most years, but visits in the first two years might find only Dipper, the middle two years could find both, and the last two years only Kingfisher.

The prevailing weather conditions for each year of the survey are summarised in Table 7 below (for full details see *SBR*s). Terms are somewhat subjective, and used relative to the county average for that time of year, to indicate the likely impact of the weather on population levels and breeding success for each species.

Table 7: Prevailing Weather Conditions during the Atlas Period

Breeding Season	Weather	Period				
		Previous Winter	April	May	June	July
1985	Temperature	Cold	Cold	Cold	Cool	Cool
	Precipitation	Lying snow	High	Average	High	Mixed
1986	Temperature	Very cold	Cold	Cold	Mixed	Mixed
	Precipitation	Low	High	High	Mixed	Mixed
1987	Temperature	Average	Cool	Cool	Cool	Cool
	Precipitation	High Lying snow	High	High	High	High
1988	Temperature	Mild	Cold	Warm	Mixed	Cool
	Precipitation	High	Average	High	Mixed	V. High
1989	Temperature	Mild	Cold	Warm	Hot	Hot
	Precipitation	Low	High	Low	Low	V. Low
1990	Temperature	Mild	Cool	Average	Average	V. Hot
	Precipitation	High	Average	Low	Average	Low

Note: "Mixed" indicates a considerable fluctuation, rather than a steady pattern.

INTERPRETING THE RESULTS

To summarise, each map must be interpreted in the light of variable fieldwork coverage, the possibility of a small number of mapping errors, and the natural fluctuations of bird populations and nest sites over the six years of fieldwork.

Even so, the Atlas aimed to show the broad distribution of each species breeding in Shropshire, and this has undoubtedly been achieved. *LS*

APPENDIX 2. Vital Statistics

Left to right the columns contain the following information:

English Name
The common name used in this work.

British Population (BrPop)
Figures taken from Appendix 4 of *Population Trends in British Breeding Birds* (Marchant *et al.* 1990). Abbreviations: M = million, K = thousand, ff = females, terrs = territories. Estimates enclosed in brackets are less precise than the others.*

Shropshire Population (ShrPop)
The population (number of pairs unless indicated otherwise) as estimated in the text of this work. Abbreviations: K = thousand, max = unlikely to exceed this figure, (+) = may be more.

Shropshire Tetrads (Tet)
The number of tetrads in which the species was recorded as confirmed or probably breeding as defined in the text and within the subject area and time period of this work.

Wildlife and Countryside Act 1981 (WCAct)
Entries indicate the presence of the species on the Schedules, the provisions of which are described below.

Schedule	Provisions
1	Species which are protected by special penalties:
1.1	At all times
1.2	During the "close" season
2	Species which may be killed or taken:
2.1	Outside the "close" season
2.2	By authorised persons at all times
3	Species which may be sold:
3.1	Alive at all times if bred and ringed in captivity
3.2	Dead at all times
3.3	Dead between 1 September to 28 February
4	Species which must be registered and ringed if kept in captivity.

EC Birds Directive 1979 (ECBD)
The European Community Birds Directive of 2/4/79 on the Conservation of Wild Birds 79/409/EEC contains Annexes of species lists. Entries here indicate the presence of the species on the Annexes, the provisions of which are described overleaf.

Annex Provisions

1   Species for which there is an expressed commitment to maintain populations.

2   Species which may be hunted:

  2.1   In all Member States.

  2.2   Only in specified Member States.

3   Trade in birds.

  3.1   Species in which legal trade is allowed.

  3.2   Species in which legal trade may be allowed within Member States.

  3.3   Species to be the subject of "Studies of their biological status" with a view to subsequent inclusion in 3.2.

The prefix "x" is used to indicate exclusion from a provision.

Berne Convention (B)

The Convention, ratified by the UK in 1983, carries an obligation to protect and conserve a wide range of flora and fauna (including their habitats), especially those listed as endangered or vulnerable. "Y" indicates the presence of the species on the Convention's list of bird species.

Red Data

Entries here indicates the presence of the species in the Schedules of *Red Data Birds in Britain* (Batten *et al.* 1990).

Schedule Provisions

1a/BI   Breeding in internationally significant numbers (>20% of the north-west Europe population).

1b/WI   Non-breeding in internationally significant numbers (>20% of the north west Europe population).

2 /BR   Rare breeder (<300 pairs).

3 /BD   Declining breeder (>50% sustained decline since 1960).

4a/BL   Localised breeder (>50% of the population in the ten most populated areas).

4b/WL   Localised non-breeder (>50% of the population in the ten most populated areas).

5 /SC   Special category — show cause for concern or declining numbers but inadequate data to quantify the extent of the problem.

CS (candidate species) indicates that the species is included in a list of possible future candidates for the *Red Data Book* (p.345) and their future status ought to be watched carefully.

Rare Birds Breeding Panel (R)

"Y" indicates the presence of the species on the RBBP's list. The RBBP collects data on all species whose British breeding population is 300 pairs or less and others whose populations fluctuate wildly or may be sensitive to disturbance. All records of breeding of these species (at whatever level) should be forwarded to the RBBP as well as to the County Bird Recorder c/o the Shropshire Ornithological Society.

ICBP Endangered Species
Only Corncrake, now probably extinct as a breeding species in Shropshire, features in the International Council for Bird Preservation's Technical Publication No 8. The ICBP *World Check List of Threatened Birds* (Collar & Andrew 1988). The designation indicates species rarity or decline on a global scale.

Notes:
1.   * indicates the need to refer to source for further detail.
2.   > means greater than, < means lesser than.

Acknowledgement
The non-Shropshire data has been taken from birdBASE, the information centre of Western Palearctic birds.   *John Tucker*

English name	BrPop	ShrPop	Tet	WCAct	ECBD	B	Red Data	R
Little Grebe	(7–13K)	250–330	83	–	–	Y	–	–
Great Crested Grebe	(3–5K)	150–200	71	–	–	–	–	–
Black–necked Grebe	26–35	1	1	1.1 4	–	Y	2/BR 4a/BL	Y
Grey Heron	9.5–9.6K	100–120	45	–	–	–	–	–
Mute Swan	3.1–3.2K	52–74	129	–	2.2(xUK)	–	–	–
Greylag Goose	(2K)	50–75	25	1.2* 2.1	2.1 3.2	–	1b/WI	–
Canada Goose	(4–5K)	550–1000	273	2.1	2.1	–	–	–
Barnacle Goose	–	1	3	–	1	Y	1b/WI 4b/WL	–
Shelduck	12K	1–3	12	–	–	Y	1b/WI 4b/WL	–
Mandarin	(2–3K)	3–6	4	–	–	–	–	–
Gadwall	600	1–2	5	2.1	2.1	–	1b/WI 4b/WL	–
Teal	3–4.5K	5	40	2.1 3.3	2.1 3.2	–	1b/WI	–
Mallard	(50–100K)	2–4K	666	2.1 3.3	2.1 3.1	–	–	–
Garganey	40–90	0	3	1.1	2.1	–	2/BR	Y
Shoveler	1–1.5K	3 max	15	2.1 3.3	2.1 3.3	–	4b/WL	–
Pochard	200–300	8 max	11	2.1 3.3	2.1 3.2	–	1b/WI 2/BR	Y
Tufted Duck	7K	160–240	190	2.1 3.3	2.1 3.2	–	–	–
Goosander	1–1.3K	5 max	5	–	2.2(xUK)	–	–	–
Ruddy Duck	(350–400)	40	40	–	–	–	–	–
Marsh Harrier	60ff	0	1	1.1	1	Y	2/BR	Y
Hen Harrier	400–500ff	0	1	1.1 4	1	Y	5/SC	–
Goshawk	100	a few	–	1.1 4	1*	–	2/BR	Y
Sparrowhawk	25K	600–1.8K	276	4	–	–	–	–
Buzzard	12–15K	300 max	302	4	–	–	CS	–
Kestrel	70K	700–1.4K	348	1.1 4	–	–	–	–
Merlin	550–650	6 max	7	1.1 4	1	–	5/SC	–
Hobby	500	8 max	12	1.1 4	–	–	–	Y
Peregrine	750–1K	2	2	1.1 4	1	Y	1a/BI	–
Red Grouse	(200–400K)	100 max	13	–	2.1 3.1	–	1a/BI	–
Red–legged Partridge	(100–200K)	2K	530	–	2.1 3.1	–	CS	–
Grey Partridge	(200–400K)	2.5K	530	–	1* 2.1 3.1	–	3/BD	–
Quail	50–600	5–8	76	1.1 4	2.2(xUK)	–	2/BR	Y
Pheasant	(3–3.5Mff)	? *	712	–	2.1 3.1	–	–	–
Water Rail	(1.5–3K)	20(+)	5	–	2.2(xUK)	–	–	–
Corncrake	550–600	0	0	1.1 4	1	Y	3/BD	Y
Moorhen	(200–225K)	3.5–7K	655	2.1	2.2	–	–	–
Coot	(40–70K)	1–2K	313	2.1 3.3	2.1 3.2	–	–	–
Oystercatcher	33–43K	6(+)	9	–	2.2(xUK)	–	1b/WI 4b/WL	–
Little Ringed Plover	610–630	10(+)	9	1.1 4	–	Y	–	–
Lapwing	200–225K	2.3K	660	–	2.2(xUK)	–	CS	–

English name	BrPop	ShrPop	Tet	WCAct	ECBD	B	Red Data	R
Snipe	30K	200–300 *	58	2.1 3.3	2.1 3.3	–	CS	–
Woodcock	(15–35Kff)	150–300	108	2.1 3.3	2.1 3.3	–	–	–
Curlew	33–38K	700	506	–	2.2	–	1a/BI 1b/WI	–
Redshank	30–33K	30–50	18	–	2.2	–	1b/WI 4b/WL	–
Common Sandpiper	17–20K	20–40	32	–	–	Y	–	–
Black–headed Gull	(150–200K)	100–200	22	–	2.2(xUK)	–	–	–
Feral Pigeon	100K	2–6K	240	2.2 3.2	–	–	–	–
Stock Dove	100K	7–10K	661	–	2.2(xUK)	–	–	–
Wood Pigeon	2.5M	35K(+)	861	2.2 3.2	2.1 3.1	–	–	–
Collared Dove	100K	2–3K	711	2.2	2.2(xUK)	–	–	–
Turtle Dove	(75–100K)	250–600	125	–	2.2(xUK)	–	CS	–
Cuckoo	(15–22Kff)	175–350	462	–	–	–	–	–
Barn Owl	4.4K	140	143	1.1 3.1	–	Y	5/SC	–
Little Owl	7–14K	600–1.8K	352	–	–	Y	–	–
Tawny Owl	50–100K	900–1.8K	434	–	–	Y	–	–
Long–eared Owl	(2–6K)	20(+)	10	–	–	Y	–	–
Nightjar	1.8–2.1K	6 max	2	–	1	Y	5/SC	–
Swift	70–80K	1400	440	–	–	–	–	–
Kingfisher	(4–6K)	140–350	137	1.1 4	1	Y	CS	–
Green Woodpecker	10–15K	500–1K	286	–	–	Y	–	–
Great Spotted Woodpecker	30–40K	1.5–3K	507	–	–	Y	–	–
Lesser Spotted Woodpecker	3–6K	250–500	108	–	–	Y	–	–
Skylark	2M	14K	601	–	2.2(xUK)	–	–	–
Sand Martin	300–500K	4K	82	–	–	Y	CS	–
Swallow	500K	3–6K	841	–	–	Y	CS	–
House Martin	(250–450K)	7K	785	–	–	Y	–	–
Tree Pipit	100K	900–1.8K	150	–	–	Y	–	–
Meadow Pipit	1–1.5M	2.5–5K	156	–	–	Y	–	–
Yellow Wagtail	(80–100K)	1.15–2.3K	234	–	–	Y	CS	–
Grey Wagtail	(17–34K)	275–500	224	–	–	Y	–	–
Pied Wagtail	300K	2.5–5K	668	–	–	Y	–	–
Dipper	(20–22K)	160–480	166	–	–	Y	CS	–
Wren	3–3.5M	100K	842	–	1*	Y	–	–
Dunnock	2M terrs.	22–26K	804	3.1	–	Y	–	–
Robin	3.5M	100K	857	–	–	Y	–	–
Nightingale	4–5K	? *	2	–	–	Y	CS	–
Black Redstart	80–110	0	2	1.1 4	–	Y	2/BR	Y
Redstart	140K	1.4–4.2K	266	–	–	Y	CS	–
Whinchat	(18–35K)	110–275	55	–	–	Y	CS	–
Stonechat	(10–20K)	25 max	23	–	–	Y	CS	–
Wheatear	(50–60K)	180–300	76	–	–	Y	CS	–
Ring Ouzel	(8–15K)	7–12	6	–	–	–	CS	–
Blackbird	4.5–5M	160–190K	866	3.1	2.2(xUK)	–	–	–
Song Thrush	1.5M	17.5–35K	719	3.1	2.2(xUK)	–	–	–
Mistle Thrush	300K	4.8–5.5K	683	–	2.2(xUK)	–	–	–
Grasshopper Warbler	(10–18K)	90–180	34	–	–	Y	–	–
Sedge Warbler	(150–200K)	250–500	76	–	–	Y	CS	–
Reed Warbler	40–80K	250–500	42	–	–	Y	–	–
Lesser Whitethroat	50K	700–1K	184	–	–	Y	–	–
Whitethroat	400–500K	5–10K	612	–	–	Y	CS	–
Garden Warbler	200K	2–2.6K	461	–	–	Y	–	–
Blackcap	800K	6.5–10K	646	–	–	Y	–	–
Wood Warbler	16–18.5K	400 males	139	–	–	Y	–	–
Chiffchaff	400–500K	3–6K	580	–	–	Y	–	–
Willow Warbler	2.5M	40K(+)	772	–	–	Y	–	–
Goldcrest	500–600K	30–50K	381	–	–	Y	–	–
Firecrest	80–100	0	1	1.1 4	–	Y	2/BR	Y
Spotted Flycatcher	200K	2K	652	–	–	Y	CS	–
Pied Flycatcher	(40K)	2K	198	–	–	Y	–	–
Long–tailed Tit	200K terrs	2.5–3.7K	625	–	–	–	–	–

188

English name	BrPop	ShrPop	Tet	WCAct	ECBD	B	Red Data	R
Marsh Tit	(120–140K)	1.75–3.5K	290	–	–	Y	–	–
Willow Tit	50–100K	1.4–2.8K	223	–	–	Y	–	–
Coal Tit	500–700K	20K	491	–	–	Y	–	–
Blue Tit	3.5M	60K	862	–	–	Y	–	–
Great Tit	2M	30K	819	–	–	Y	–	–
Nuthatch	50K	2.25–4.5K	452	–	–	Y	–	–
Treecreeper	200–250K	5–10K	488	–	–	Y	–	–
Jay	(80K)	1.6K	484	2.2 3.1	–	–	–	–
Magpie	(300–400K)	15–20K	846	2.2 3.1	–	–	–	–
Jackdaw	(350–400K)	10K	762	2.2 3.1	–	–	–	–
Rook	850–860K	25K	593	2.2	–	–	–	–
Carrion Crow	(800K–1M)	17.5–22K	839	2.2	–	–	–	–
Raven	(3.5–4K)	30–35	99	–	–	–	CS	–
Starling	(3–5M)	27–54K	838	2.2 3.1	–	–	–	–
House Sparrow	(4–4.5M)	70K	845	2.2	–	–	–	–
Tree Sparrow	(260K)	5K	394	–	–	–	CS	–
Chaffinch	5M	140K	856	3.1	–	–	–	–
Greenfinch	800K	9–10K	658	3.1	–	Y	–	–
Goldfinch	250–300K	4.5K	684	3.1	–	Y	–	–
Siskin	(14–28K)	100	23	3.1	–	Y	–	–
Linnet	(600–700K)	2–4K	570	3.1	–	Y	CS	–
Redpoll	140–150K	100 max	40	3.1	–	Y	–	–
Crossbill	500–5000	? *	10	1.1 4	–	–	–	–
Bullfinch	300–350K	1.5–3K	576	3.1	–	–	–	–
Hawfinch	5–10K	110–220 max	11	–	–	Y	–	–
Yellowhammer	1.5M	35K	810	3.1	–	Y	–	–
Reed Bunting	400K	570–1900	190	3.1	–	Y	–	–
Corn Bunting	30K	400–900	130	–	–	–	CS	–

APPENDIX 3. References

Allin, E.K. (1968). Breeding Notes on Ravens in North Wales. *British Birds* 61:541–545.

Anon. (1938). Waterfowl at Walcot. *Avicultural Magazine* 80:132–135.

Batten, L.A., Bibby, C.J., Clement, P., Elliott, G.D. & Porter, R.F. (1990). *Red Data Birds in Britain.* Poyser, London.

Beckwith, W.E. (1879). List of Shropshire Birds. *Trans. of Shropshire Archaeological Soc.* 2:381.

Bibby, C.J. (1989). A Survey of Breeding Wood Warblers in Britain, 1984–85. *Bird Study* 36: 56–72.

Campbell, B. (1960). The Mute Swan Census in England and Wales, 1955/56. *Bird Study* 7:208–223.

Collar, N.J. & Andrews, P. (1988) *Birds to Watch: The ICBP World Check List of Threatened Birds.* Cambridge/ICBP.

Coombs, C.F.J. (1978). *The Crows.* Batsford, London.

Cramp, S., Simmons, K.E.L. et al. (eds) (1977 onwards). *The Birds of the Western Palearctic* (7 vols.). Oxford University Press, Oxford.

Dare, P.J. (1986). Raven Populations in Two Upland Regions of North Wales. *Bird Study* 33:179–189.

Eltringham, S.K. (1963). The British Population of the Mute Swan. *Bird Study* 10:10–28.

Eyton, T.C. (1838). An Attempt to Ascertain the Fauna of Shropshire & North Wales. *Annals and Magazine of Natural History.*

Forrest, H.E. (1899). *The Fauna of Shropshire.*

Forrest, H.E. (1908). Vertebrates of Shropshire. *Victoria County History of Shropshire.* Vol. 1.

Forrest, H.E. (1909). Record of Bare Facts. *Caradoc & Severn Valley Field Club* 19:23 & 29.

Forrest, H.E. (1918). Ravens Nesting again in Shropshire. *British Birds* 12:19.

Forrest, H.E. (1930). Record of Bare Facts. *Caradoc & Severn Valley Field Club* 40:13.

Forrest, H.E. (1934). Record of Bare Facts. *Caradoc & Severn Valley Field Club* 44:12.

Fox, A.D. (1988). Breeding Status of the Gadwall in Britain and Ireland. *British Birds* 81:51–66.

Fox, A.D. (1991). History of the Pochard Breeding in Britain. *British Birds* 84:83–98.

Hardy, E. (1970). The Return of the Peregrine. *Shropshire Magazine,* September:23.

Hardy, E. (1971). Falcons of the Hills. *Shropshire Magazine,* September:21.

Harrison, G. & Sankey, J. (1987) *Where to Watch Birds in the West Midlands.* Helm, London.

Hudson, R. (1976). Ruddy Ducks in Britain. *British Birds* 69:132–143.

Jepson, P. (1991). *Shrewsbury Countryside Strategy.* Shrewsbury & Atcham Borough Council.

Lack, P. (ed.) (1986). T*he Atlas of Wintering Birds in Britain and Ireland*. Poyser, Calton.

Little, B. & Furness, R.W. (1985). Long-distance Moult Migration by British Goosanders. *Ringing and Migration* 6:77–82.

Marchant, J.H., Hudson, R., Carter, S.P & Whittington, P.A. (1990). *Population Trends in British Breeding Birds*. BTO/NCC, Tring.

Mikkola, H. (1983). *Owls of Europe*. Poyser, Calton.

Ministry of Agriculture, Fisheries & Food. (1988). *MAFF in the Midlands and Western Region*. MAFF, Wolverhampton.

Murton, R.K., Westwood, N.J. & Isaacson, A.J. (1964). The feeding habits of the Woodpigeon, Stock Dove and Turtle Dove. *Ibis* 106:174–188.

Newton, I. (1986). *The Sparrrowhawk*. Poyser, Calton.

O'Connor, R.J. & Shrubb, M. (1986) *Farming and Birds*. Cambridge University Press, Cambridge.

Ogilvie, M.A. (1975). *Ducks of Britain and Europe*. Poyser, Berkhamsted.

Ormerod, S.J. & Tyler, S.J. (1987). Aspects of the Breeding Ecology of Welsh Grey Wagtails. *Bird Study* 34:43–51.

Ormerod, S.J. & Tyler, S.J. (1990). Population characteristics of Dipper roosts in Mid and South Wales. *Bird Study* 37:165–170.

Ormerod, S.J., Tyler, S.J., Pester, S.J. & Cross, N.V. (1988). Censusing the distribution and population of birds along upland rivers using measured ringing effort: a preliminary study. *Ringing and Migration* 9:71–82.

Osborne, P. (1982). Some Effects of Dutch Elm Disease on Nesting Farmland Birds. *Bird Study* 29:2–16.

Owen, M., Atkinson-Willes, G.L. & Salmon, D.G. (1986). *Wildfowl in Great Britain* (Second Edition). Cambridge University Press, Cambridge.

Parslow, J.L.F. (1967). Changes in Status among Breeding Birds in Britain & Ireland. *British Birds* 60:2–4.

Petty, S.J. (1989). Goshawks: their Status, Requirements and Management. *Forestry Commission Bulletin* 81. HMSO, London.

Potts, G.R. (1989). The Impact of releasing Hybrid Partridges on Wild Red-legged Populations. *Game Conserv. Ann. Review* 20:81–85.

Ratcliffe, D. (1980). *The Peregrine Falcon*. Poyser, Calton.

Reynolds, C.S. (1979). The Limnology of the Eutrophic Meres of the Shropshire-Cheshire Plain. *Field Studies* 5:93–173.

Robertson, H.A. (1990). Breeding of Collared Doves in rural Oxfordshire. *Bird Study* 37:73–83.

Rowley, T. (1972). *The Shropshire Landscape*. Hodder & Stoughton.

Royal Society for the Protection of Birds (1991). *Death by Design (The Persecution of Birds of Prey and Owls in the U.K. 1979–89)*. R.S.P.B., Sandy.

Rutter, E.M., Gribble, F.C. & Pemberton, T.W. (1964). *A Handlist of the Birds of Shropshire*. Shropshire Ornithological Society, Shrewsbury.

Sage, B.L. & Vernon, J.D.R. (1978). The 1975 National Survey of Rookeries. *Bird Study* 25:64–86.

Sharrock, J.T.R. (ed.) (1976). *The Atlas of Breeding Birds in Britain and Ireland*. Poyser, Berkhamsted.

Shawyer, C.R. (1987). *The Barn Owl in the British Isles: its Status Present and Future*. Hawk Trust, London.

Shropshire Wildlife Trust (1989). *Losing Ground in Shropshire*. S.W.T. Shrewsbury.

Sinker, C.A., Packham, J.R., Trueman, I.C., Oswald, P.H., Perring, F.H. & Prestwood, W.V. (1985). *Ecological Flora of the Shropshire Region*. Shropshire Trust for Nature Conservation.

Spencer, R. & the Rare Breeding Birds Panel. (1990). Rare Breeding Birds in the United Kingdom in 1988. *British Birds* 83:353–390.

Stroud, D.A. & Glue, D. (1991). *Britain's birds in 1989/90: The Conservation and Monitoring Review*. BTO/NCC, Thetford.

Tyler, S.J., Ormerod, S.J. & Lewis, J.M.S. (1990). The post-natal and breeding dispersal of Welsh Dippers. *Bird Study* 37:18–23.

Wassell, D. (ed.) (1981). *Wildlife in Telford*. Telford Development Corporation.

APPENDIX 4. Plants and Animals
List of plants and animals mentioned in the text

PLANTS

alder	*Alnus glutinosa*
ash	*Fraxinus excelsior*
beech	*Fagus sylvatica*
bilberry	*Vaccinium myrtillus*
birch	*Betula spp.*
blackthorn	*Prunus spinosa*
bracken	*Pteridium aquilinum*
bramble	*Rubus spp.*
common fumitory	*Fumaria officinalis*
common heather	*Calluna vulgaris*
common reed	*Phragmites australis*
elm	*Ulmus spp.*
gorse	*Ulex spp.*
hawthorn	*Crataegus monogyna*
hazel	*Corylus avellana*
heather	*Calluna vulgaris* or *Erica spp.*
holly	*Ilex aquifolium*
hornbeam	*Carpinus betulus*
Italian ryegrass	*Lolium multiflorum*
ivy	*Hedera helix*
larch	*Larix spp.*
nettle	*Urtica spp.*
oak	*Quercus spp.*
reed canary grass	*Phalaris arundinacea*
Scots pine	*Pinus sylvestris*
Sitka spruce	*Picea sitchensis*
sycamore	*Acer pseudoplatanus*
water crowfoot	*Ranunculus aquatilis*
wild cherry	*Prunus avium*
willow	*Salix spp.*
yew	*Taxus baccata*

ANIMALS

caddis fly	*Trichoptera*
common shrew	*Sorex araneus*
field vole	*Microtus agrestis*
fox	*Vulpes vulpes*
grey squirrel	*Sciurus carolinensis*
mayfly	*Ephemeroptera*
mink	*Mustela vison*
rabbit	*Oryctolagus cuniculus*
stoat	*Mustela ermina*
stonefly	*Plecoptera*
weasel	*Mustela nivalis*
wood ant	*Formica rufa*
wood mouse	*Apodemus sylvaticus*

APPENDIX 5. Place Names
List of place names mentioned in the text with the four figure national grid reference

Acton pool	SO 3184	Corve, River	SO 5691	Llyn Rhuddwyn	SJ 2328
Alkmond Park Pool	SJ 4716	Cound Brook	SJ 5006	Llynclys Hill	SJ 2723
Allscott Sugar Factory	SJ 6012	Cranmere Bog	SO 7597	Long Mountain	SJ 2707
Anchor	SO 1785	Craven Arms	SO 4382	Long Mynd, The	SO 4194
Ape Dale	SO 4990	Cronkhill	SJ 5308	Loton Park	SJ 3514
Apley Castle	SJ 6513	Crose Mere	SJ 4330	Lower Wood	SO 4697
Aqualate	SJ 7720	Crosemere	SJ 4329	Ludlow	SO 5175
Ashes Hollow	SO 4392	Crudgington	SJ 6317	Lydbury North	SO 3586
Ashford Carbonell	SO 5270	Culmington	SO 4982	Lyneal Wood	SJ 4531
Astbury	SO 7289	Dowles	SO 7776	Market Drayton	SJ 6734
Atcham	SJ 5409	Dudmaston (Hall)	SO 7488	Marton Pool	SJ 4423
Baggy Moor	SJ 3827	Earl's Hill	SJ 4004	Mason's Bank	SO 2287
Beckbury	SJ 7601	East Onny, River	SO 3996	Melverley	SJ 3316
Benthall Edge	SJ 6603	Edgerley	SJ 3518	Middletown	SJ 3012
Berrington Pool	SJ 5207	Edgmond	SJ 7219	Middleton Hill	SJ 3013
Betton Pool	SJ 5107	Ellesmere	SJ 4034	Millichope Park	SO 5288
Bishop's Castle	SO 3288	English Frankton	SJ 4529	Mor Brook	SO 6694
Black Bank	SO 2387	Ercall, The	SJ 6409	Much Wenlock	SO 6299
Black Mountain	SO 1983	Farley Brook	SJ 6302	Mucklewick Hill	SO 3397
Black Rhadley Hill	SO 3495	Felindre	SO 1781	Nedge Hill	SJ 7107
Bog, The	SO 3597	Felton Butler	SJ 3917	Nesscliffe	SJ 3819
Borle Brook	SO 7186	Fenemere	SJ 4422	Newcastle	SO 2482
Boyne Water	SO 5984	Fenn's Moss	SJ 4836	Newport	SJ 7418
Brick Kiln Farm	SJ 5543	Frodesley	SJ 5101	Newton Mere	SJ 4234
Bridgnorth	SO 7192	Gatten wood	SO 3798	Nib Heath	SJ 4118
Brompton	SJ 5407	Grinshill	SJ 5223	Noneley	SJ 4727
Broseley	SJ 6701	Grit Hill	SO 3398	Norbury Hill	SO 3594
Brown Clee	SO 5986	Haddon Hill	SO 4395	Norbury pool	SO 3692
Brown Moss	SJ 5639	Halston (Hall)	SJ 3431	Northwood	SJ 4633
Bryn Shop	SO 1783	Hanmer Mere	SJ 4539	Onny, River	SO 3987
Bucknell	SO 3573	Harley Brook	SJ 5900	Onslow	SJ 4312
Buildwas	SJ 6404	Hatton Grange	SJ 7604	Oswestry	SJ 2929
Bury Ditches	SO 3283	Haughmond Hill	SJ 5414	Oxon Pool	SJ 4513
Caer Caradoc	SO 4795	Hawkstone Park	SJ 5729	Park Hall Camp	SJ 3031
Camlad, River	SO 3092	Heath Mynd	SO 3393	Perry, River	SJ 3926
Cardingmill Valley	SO 4494	Hencott Pool	SJ 4916	Pim Hill	SJ 4821
Catherton Common	SO 6378	Hilton	SO 7795	Pole Cottage	SO 4193
Cefn Coch	SJ 2333	Hoar Edge	SO 6077	Polmere	SJ 4109
Cefn Gunthly	SO 3395	Hodnet Heath	SJ 6126	Prees	SJ 5533
Cefn Hepreas	SO 2678	Hope Valley	SJ 3300	Prees Heath	SJ 5637
Ceiriog, River	SJ 2438	Hopesay	SO 3983	Priorslee Flash	SJ 7110
Chelmarsh Reservoir	SO 7387	Hopton Cangeford	SO 5480	Priorslee Lake	SJ 7109
Chetwynd pool	SJ 7420	Ironbridge	SJ 6703	Quatford	SO 7390
Chirbury	SO 2698	Isombridge	SJ 6013	Ragleth Hill	SO 4592
Clive	SJ 5124	Knighton	SO 2872	Rea Brook	SJ 3404
Clun Forest	SO 2286	Knighton Resr.	SJ 7328	Rea, River	SO 6773
Clun, River	SO 3381	Lawley, The	SO 4997	Rea valley	SJ 3404
Clun	SO 3080	Leebotwood	SO 4798	Redlake, River	SO 3275
Clungunford	SO 3978	Leighton	SJ 6105	Red Wood	SO 3183
Coalport	SJ 6902	Lilleshall hill	SJ 7315	Rednal	SJ 3628
Cole Mere	SJ 4333	Lilleshall (Hall)	SJ 7414	Riddings	SO 1986
Colemere	SJ 4332	Linley Big Wood	SO 3494	River Camlad	SO 3092
Combermere Park	SJ 5844	Linley Hill	SO 3594	River Ceiriog	SJ 2438
Condover (gravel pit)	SJ 4804	Llanymynech	SJ 2620	River Clun	SO 3381
Corve Dale	SO 5387	Lloyds Coppice	SJ 6903	River Corve	SO 5691

River East Onny	SO 3996	Sleap	SJ 4826	Tyrley Locks	SJ 6932
River Onny	SO 3987	Sowdley Wood	SO 3280	Vennington	SJ 3309
River Perry	SJ 3926	Stanmore	SO 7492	Venus Pool	SJ 5406
River Rea	SO 6773	Stanton Lacy	SO 4978	Vyrnwy, River	SJ 3216
River Redlake	SO 3275	Stapeley Hill	SO 3199	Walcot Park	SO 3485
River Roden	SJ 5624	Steel Heath	SJ 5436	Weald Moors, The	SJ 6715
River Severn	SJ 4014	Stiperstones	SO 3698	Wellington	SJ 6411
River Tanat	SJ 2421	Stockton Wood	SJ 2601	Welsh Frankton	SJ 3632
River Teme	SO 2674	Stow Hill	SO 3074	Wem Moss	SJ 4734
River Tern	SJ 6229	Sutton Maddock	SJ 7201	Wenlock Edge	SO 5089
River Vyrnwy	SJ 3216	Swan Farm mine	SJ 6406	Westbury	SJ 3509
River West Onny	SO 3494	Telford	SJ 6909	West Onny, River	SO 3494
River Worfe	SO 7594	Tanat, River	SJ 2421	Whitchurch	SJ 5441
Rock, The	SJ 6809	Teme, River	SO 2674	Whitcliffe	SO 5074
Roden, River	SJ 5624	Tern, River	SJ 6229	Whittington Castle	SJ 3231
Rose Grove	SO 1885	The Bog	SO 3597	Whixall Moss	SJ 4935
Severn, River	SJ 4014	The Ercall	SJ 6409	Willey Park	SO 6699
Shavington (Hall)	SJ 6338	The Lawley	SO 4997	Wood Lane (gravel pit)	SJ 4232
Shelve	SO 3399	The Long Mynd	SO 4194	Worfe, River	SO 7594
Shelve Pool	SO 3397	The Rock	SJ 6809	Worfield	SO 7595
Shifnal	SJ 7507	The Weald Moors	SJ 6715	Worthen	SJ 3204
Shirlett	SO 6597	The Wrekin	SJ 6208	Wrekin, The	SJ 6208
Shrawardine	SJ 3915	Titterhill	SO 3577	Wroxeter	SJ 5608
Shrewsbury	SJ 4912	Titterstone Clee	SO 5978	Wyre Forest	SO 7576
Shrewsbury Sewage Farm	SJ 5213	Tong Lake	SJ 7907		

APPENDIX 6. Sponsors and Supporters

The generous support of members and others has been mentioned elsewhere but this appendix lists those who gave to the Atlas Appeals. Initially funds were raised to help with the costs of the computer programme then, when the fieldwork was finished, a second appeal was launched to provide funds to enable the Atlas to be published at a reasonable price. The success of the appeals enabled English Nature to make a generous grant towards publication costs, and without this support the Atlas would not have been published in this format. The Countryside Trust has financed the recent appeal and the promotion costs.

Albrighton District Conservation Society, The Belvidere School Shrewsbury, Bridgnorth District Council, British Birds, Council for the Preservation of Rural England (Shrewsbury Branch), CJ Wildbird Foods Ltd, The Countryside Trust, English Nature, Field Studies Council, North Shropshire District Council, Shrewsbury & Atcham Borough Council, Shrewsbury 6th Form College, Shrewsbury Young Farmers, S.O.S. Ludlow Branch, Telford Development Corporation and The Women's Institute.

Mrs D.L. Andrew	Mr M.R. Hinks	Mr J.C. Smallwood
Mrs N.H. Barnett	Mr J. Hoadley	Mr J.M. Smith
Mr A.P. Bell	Miss J. Hocson	Mr L.N. Smith
Mr H.J. Blofield	Mrs J. Jackson	Mr M.G. Smith
Mr R. Bromley	Mr R.A. Keyse	Ms A. Suffolk
Mr I.A.R. Brown	Mr P. Lander	Mr F.J. Thomas
Mrs B.E. Burns	Mr M. Law	Mr G. Thomas
Mr E.T.B. Butcher	Miss R. Lees	Mrs K.M. Unitt
Mrs W.W. Crosthwaite	Mrs E.M. Lloyd	Lady Wakeman
Miss M. Crow	Mr R. Mayall	Mr M.J. Wall
Mr M.G. Davies	Dr J.D. McCarter	Mr M. Wallace
Mr & Mrs P.G. Deans	Mr & Mrs D. & R. Moore	Mrs B.K. Ward
Mrs M. Donoghue	Mr & Mrs J.T. Morris	Mr A. Wellings
Mr & Mrs K.D. Dowley	Mr A.D. Owen	Mr C.J. Whittles
Mr T.W. Edwards	Mr A.R. Pyper	Mrs D. Williams
Mrs V. Evans	Dr J.C. Ryle	Mr & Mrs R. Wood
Mrs G.M. Francis	Mr & Mrs F.C. Salter	Mr D.C. Woodger
Miss M. Fuller	Mr J. Sankey	Mr & Mrs C.E. Wright
Mrs M. Hegarty	Mrs M. Saunders	
Mr P.W. Hinde	Mr A. Shimeld	

APPENDIX 7. Acknowledgements

In addition to the Observers, Sponsors and Supporters listed elsewhere, this Atlas would not have been possible without the commitment and hard work of the people listed below:–

The landowners and farmers of Shropshire who gave us access to their land.

The Committee of the Shropshire Ornithological Society who authorised the project and gave moral support throughout.

Atlas Sub-Committee
Glenn Bishton (1984–1985), Peter Deans, Jack Sankey, Leo Smith (1986–1992), John Tucker, and Chris Whittles who attended all or some of the 40 sub-committee meetings with some of the later ones closing after midnight!

Area Co-ordinators (ACO's)
ACO's who recruited and briefed fieldworkers, and persuaded them to cover new tetrads each year; transcribed all records onto master cards, one for each tetrad; and checked the computer printouts against the master cards:–

1. North-west — Allan Dawes.
2. North — Derek Evans (1985) Derek Sparkes (1986–1990).
3. North-east — Gerry Thomas.
4. West — Michael Wallace.
5. North-central — Colin Wright.
6. East — Glenn Bishton (1985) Leo Smith (1986–1990).
7. South-west — Ian Rowat (1985) Tom Wall (1986–1990).
8. South-central — Nigel Green.
9. South-east — Brian Jones.
10. South — John Milner.

Jack Sankey (County Bird Recorder 1984–1990) who handled all the general enquiries and maintained the confidential records.

Bryonie Wall, Gwen Wallace, Eileen Thomas, Pat Wright and Rosemary Moore who helped ACO's with data checking or species accounts.

The families of all the ACO's who suffered neglect especially during the annual peaks of data gathering.

Computer Program TETRAD™
Martin Humphries who was specially commissioned to write the computer program to store, process and map the data that had been gathered. TETRAD™ was subseqently developed and marketed through natureBASE, Holbrook House, 105 Rose Hill, Oxford OX4 4HT, (0865) 770490 or Fax (0865) 775656.

Data Input
Jean Hooson deserves special thanks for loading the vast majority of the data into the computer, with some assistance from Rosemary Moore, Dawn Balmer and John Fox.

The Artists
Jack Sankey, Geoff Hall, John Martin and Philip Walker, who gave their time and expertise free.

The Species Account writers
Tony Cross, Allan Dawes, John Milner, Jack Sankey, Leo Smith, Derek Sparkes, Gerry Thomas, Tom Wall and Michael Wallace who wrote the various species accounts and accepted the cuts and changes forced upon them by the editor as the accounts were checked, questioned and fitted into place.

Office and Computer facilities
Provided by Shropshire Wildlife Trust at their Headquarters, initially at St George's School and subsequently at 167 Frankwell, Shrewsbury SY3 8LG, with the support of John Tucker, Head of Conservation, who acted as liaison officer.

Fundraising
Sponsorship and grant applications were handled by Leo Smith. The Appeal leaflet was designed by Andrew White. Peter Deans led the Atlas Appeal to the membership of the SOS. Chris Whittles, as treasurer, tried to keep our expenses under control, handled the money and paid the bills.

Typing copy for the printer
Illa Patel and Janet Smith.

Editorial work
Drawings — Jack Sankey. Text — Leo Smith.
Research and tetrad map of Shropshire — Steve Shirra.

Map Design
Peter Tucker of PGT Design, Holbrook House, 105 Rose Hill, Oxford OX4 4HT, (0865 770490), prepared the map data from TETRAD™ and other sources for use in this Atlas. Special thanks to the cartography unit of MAFF in Wolverhampton for supplying original artwork for Maps 3, 6 & 7. Also to Hodder & Stoughton for permission to copy the Natural Regions of Shropshire map from *The Shropshire Landscape*, the Shropshire Wildlife Trust for permission to use data from *The Ecological Flora of the Shropshire Region* to prepare the altitude maps and the distribution map of Common Reed, and Shropshire County Council for permission to use the original Forestry and Woodland map to produce Map 8.

Content
Extensive comment on first drafts of all species accounts by Leo Smith. Subsequent drafts of these accounts and all other content was commented on by Peter Deans,

Jack Sankey, Leo Smith, Colin Wright and John Tucker. Special thanks to Anne Edwards and Mike Fitch who also read all the text and made valuable comments. Synthesis, co-ordination and liaison with writers by Leo Smith.

Design and Printing
This book was designed and printed by Livesey Limited of 101 Longden Road, Shrewsbury SY3 9EB, (0743) 235651. The help and patience of the staff at Livesey's, especially their designer, Mike Bowring, is gratefully acknowledged.

Proofreading
Mike Fitch, who also gave technical advice on style, format and presentation, Anne Edwards, Dawn Balmer, Lena Dunkley, Brenda Faulconbridge, Jean Hooson, Michael Law, Ann Mercer and Janet Williams.

Special Thanks
To Glenn Bishton, who in the beginning persuaded everyone to get started (and then moved to Devon!). Finally, extra special thanks to Anne Suffolk and Pat Wright, who suffered with good grace whilst the Atlas gradually took over their homes and their partners.

Colin Wright
Chairman of the Atlas Sub-Committee

APPENDIX 8. Observers

List of Observers who contributed records

Allen, W. D.
Allwood, J.
Archibald, K.
Armstrong, J.
Arrowsmith, J. W.
Atkinson, R.
Austin, G.
Baggley, I.
Bailey, M.
Baker, M. D.
Ball, M.
Balmer, D.
Bardiger, M.
Barnes, J.
Barnett, N. H.
Baron, T.
Barrow, D.
Bartlett, K.
Bayworth, M.
Beard, F.
Bell, A. P.
Bennett-Lloyd, P. T.
Bennett, P.
Bennett, R.
Beresford, B.
Bingham, J.
Bishton, G.
Blakeway, G.
Blofield, H. J.
Bluck, R.
Blunt, A. G.
Boniface, D. L.
Boston, B.
Bowen, A. L.
Box, J.
Bradney, R. J.
Brisbourne, L.
Burns, B.
Burrell, W. D.
Butter, A.
Byles, A. J.
Cameron, C.
Campbell, J.
Challinor, P.

Chambers, E.
Chambers, M. R.
Chapman, K. A.
Chapman, M.
Cherrington, J.
Chester-Master, R. H.
Chesterman, D. K.
Clarke, J.
Clarke, M. F. L.
Clay, B.
Clutterbuck, L. E.
Coates, S.
Cole, G. V.
Conde, S. R.
Conneely, J.
Cooksey, G.
Cooper, S. M.
Corfield, R. A.
Cotgreave, P.
Cotgreave, R. E.
Cox, K.
Crane, V.
Crawford, A.
Cross, A.
Cruickshanks, L.
Davidson, P. S.
Davies, A.
Davies, K.
Davies, P.
Davies, T.
Davis, A. K.
Davis, K.
Dawes, A. P.
Dawes, R. A.
Deans, P.
Delafield, A.
Dickson, F.
Dodd, S. P.
Dodgson, J.
Dodwell, W. R. B.
Donnelly, B.
Donoghue, M.
Dowen, R. A.
Dowley, K.

Draper, B.
Driscoll, D.
Driscoll, E.
Dunkley, A. H.
Dye, R. B.
Eaton, J.
Edbrooke, S.
Edgenton, P.
Edmond, R.
Edwards, A.
Edwards, J.
Edwards, T. W.
Edwards, V.
Evans-Fisher, I.
Evanson, B. M.
Evans, D. J.
Evans, L.
Evans, M. I.
Evans, M.
Evans, V.
Farr, R.
Fathers, S.
Faulconbridge, B.
Fisher, B. M.
Fitch, M.
Flynn, G.
Ford, A. M.
Forrester, V.
Francis, M.
Franks, J. P. L.
Fuller, M.
Garbutt, D. E.
Gaunt, A.
Gibbs, R.
Gissing, J. G.
Gladwell, P.
Graham, P.
Grant, M. G.
Greenall, L.
Green, N.
Grindley, M.
Groom, A.
Hadfield, A.
Hajdasz, J.

Hall, G.	Jackson, C. E.	Mileto, R.
Hampson, B.	Jackson, P.	Milner, J.
Hampson, D. H.	Jackson, R.	Moore, R.
Hand, H.	Jardine, D. C.	Moreton, J.
Hankinson, I.	Jeavons, P.	Morgan, J.
Hanmer, J.	Jefferson, R.	Morley, J.
Harris, D.	Jepson, P.	Morris, A.
Harris, L.	Jones, B. A.	Morris, H. M.
Harris, M.	Jones, D. P.	Morrison, A.
Harrison, A.	Jones, G.	Munro, I.
Harrison, C. J.	Jones, J. A.	Murphy, A.
Harrison, J.	Jones, J. B.	Murphy, D.
Hartley, M.	Jones, J. I.	Mycock, R. J.
Harvey, F. J.	Jones, S.	Nelson, S.
Hatherway, H.	Jordan, P.	Nicholls, A. D.
Haugh, B.	Jukes, A.	Nicholls, A. T.
Hawkins, J.	July, M.	Nock, R.
Haycox, S. L.	Keyse, R. A.	Nunn, J.
Hay, J.	Killiner, D. M.	Oliver, D.
Hay, S.	Kinrade, R. P.	Onions, D.
Hazlewood, D. M.	Lander, P.	Onions, L.
Head, R.	Landucci, I. E.	Owen, E.
Heath, A.	Lane, E. S.	Pacult, N.
Hedge, N.	Langford, A. F.	Parker, J. M.
Henderson, E.	Langford, J. M.	Parker, P.
Hewitt, S.	Latham, I. A.	Parker, R.
Heywood, J. R.	Lawrence, J.	Parkinson, C.
Hinde, P. W.	Lees, R.	Parkinson, C.
Hinks, R.	Leighton, M.	Parkinson, D.
Holden, M.	Lennox, G.	Parry-Jones, D.
Holland, C.	Lennox, S.	Parsons, R. K.
Hollis, G.	Lloyd, J. B. F.	Patton, S.
Holloway, S.	Lloyd, K.	Paul, M.
Hooson, J.	Lowe, D. R.	Payton, S. M.
Howells, R.	Lucas, D.	Pedlow, D.
Howells, R. L.	Lucas, M.	Pedlow, J.
Howse, C. M.	Lumsden, K. C.	Perkins, L.
Howse, P. C.	Lynch, M.	Perrott, B.
Hudson, P.	Macguire, J.	Perrott, J.
Hudson, S.	Machin, N.	Petford, R.
Hughes, J.	Maksymiw, P.	Pollard, M. J.
Hughes, W.	Marchant, J. H.	Poole, A.
Hugh, N.	Martin, J. P.	Poole, M. H.
Hulme, V.	Martin, T.	Powell, R.
Hurrell, C.	Mason, P. C. N.	Price, C.
Ing, M. J.	May, E.	Prince, H.
Ingleston, P.	McDermott, J.	Pyatt, A.
Ingleton, L.	McPhail, P.	Pyper, A. R.
Ingleton, R.	McVey, M.	Rae, A.
Isherwood, M.	Messer, G.	Ramsay, H.

Rees, H.
Remfrey, A.
Richardson, A.
Richardson, M. D.
Richardson, R. R.
Richards, R.
Roberts, D.
Roberts, J.
Robinson, C.
Robinson, J. K.
Robinson, M.
Rogers, H. B.
Rook, G.
Rowat, I.
Russon, G.
Samways, J.
Sankey, J.
Scott, D.
Scott, D. W.
Scott, J.
Sheldon, S.
Shepherd, K.
Shirra, S.
Simkin, D.
Simpson, D.
Simpson, P.
Smallwood, B. E.
Smallwood, J.
Smith, C. A.
Smith, G.
Smith, J. M.
Smith, L.
Smith, M. G.
Smith, M.
Smith, M. S.

Smith, R.
Smith, R.
Snell, P.
Sparkes, D.
Spence, B.
Spence, G.
Stafford, S.
Staines, L.
Stanley, E.
Starr-Martin, C. C.
Stockton, B.
Suffolk, A.
Suffolk, K. E.
Swales, P. R.
Sweetman, F. S.
Swindells, R.
Swindells, S.
Talbot, D. W.
Tarrant, A. K.
Taylor, V.
Ten Hoeve, P.
Tetlow, P.
Thomas, A. R.
Thomas, C.
Thomas, G.
Thomas, S.
Thompson, I. S.
Toms, C.
Townshend, R. H.
Travis, C.
Tucker, E. A.
Tucker, J. J.
Turtle, C.
Underhill, K. J.
Unitt, K. M.

Vanderhook, J.
Vickers, D.
Wainwright, M.
Walker, C. J.
Walker, P. V.
Wall, M.
Wall, T.
Wallace, M. F.
Watson, G.
Weatherley, D.
Webb, W. J. & family
Webster, H.
Wedge, W. A.
White, G.
Whittles, C. J.
Williams, D.
Williams, G. D.
Williams, J.
Williams, K. D.
Williams, M.
Williams, P. R.
Williams, S. E.
Williamson, P.
Wingfield, J.
Wistow, R.
Wood, D.
Wood, R.
Wood, V.
Woodford, D. E.
Wright, C. E.
Wright, F.
Wright, G. F.
Wright, M.
Wyllie, J.
Young, G. E.

Records were also submitted by the Forestry Commission, Grange Farm Watch Group, the Nature Conservancy Council, and the Telford Park Rangers.

Every effort has been made to ensure that this list is complete but if we have left you off please accept our apologies (Eds.).

INDEX

Figures in **bold type** indicate the main species account and map. Appendix 2 — Vital Statistics, is not included in this index but lists all the breeding species.

Whitchurch

Ellesmere

Oswestry

Market Drayton

R. Roden

R. Tern

R. Perry

Newport

SHREWSBURY

TELFORD

Westbury

Rea Brook

Cound Brook

R. Severn

Chirbury

Church Stretton

Bridgnorth

Bishop's Castle

R. Corve

R. Onny

R. Rea

Clun

R. Teme

Ludlow

Knighton

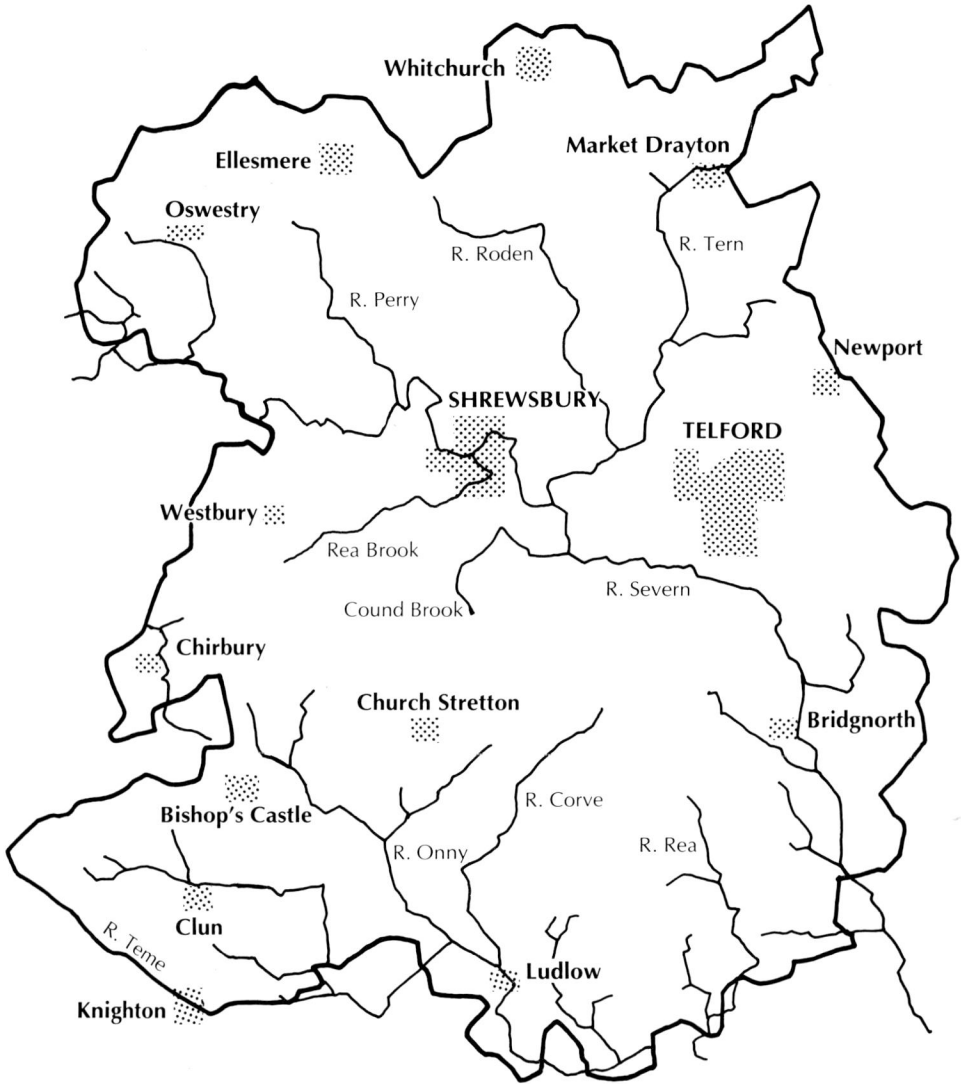

Whitchurch

Ellesmere

Oswestry

Market Drayton

R. Roden

R. Tern

R. Perry

Newport

SHREWSBURY

TELFORD

Westbury

Rea Brook

Cound Brook

R. Severn

Chirbury

Church Stretton

Bridgnorth

Bishop's Castle

R. Corve

R. Rea

R. Onny

R. Teme

Clun

Ludlow

Knighton